From Cosmic Black Hole
To Cosmo-Universe

From Cosmic Black Hole To Cosmo-Universe

An Upheaval In Physics

In an instant something happened
that changed the cosmos forever

Were Newton and Einstein both wrong?
Is relativity dead?
Is the speed of light actually a variable?
What do atoms and other quantum particles really consist of?
Did you know that in the universe there exists a large quantity
of stable anti-matter? Where?

Unveiling realms beyond the Boson and the
Universe

THE HIGGS BOSON and GRAVITY; THE GOD PARTICLE;
THE QUANTUM BIRACIAL BALANCE THEORY
&
THE THEORY OF EVERYTHING

Russell Bonney

ISBN: Softcover   978-1-4535-6855-2
             Ebook   978-1-4535-6631-2

To order additional copies of this book, contact:
Xlibris Corporation
1-800-618-969
www.xlibris.com.au
Orders@xlibris.com.au
500375

Contents

Constructing a novel unification theory for cosmological, classical and quantum physics. © *Copyright:—July 2010: Russell K. Bonney*

Centuries old scientific theories are hanging around the neck of science like a dead Albatross.
It's time for progressive science to pave the way for futuristic technologies.

THE UNIVERSE IS HEADED TOWARD ENTROPY, WHICH IS CURRENTLY BEING DELAYED BY A TENTATIVE STATE OF EQUILIBRIUM.

There was movement at the station for the word had got around . . . (Banjo Patterson)

Quantum theory might emerge from a deeper level of non weird physics . . . (Tim Collins)

Mathematics demonstrates that small changes to simple algorithms can lead to extreme complexity. It also declares that without algorithms chaos can only lead to more chaos.

PREFACE

The following will contain many statements and ascertains which may appear to be faulty physics. These may seem puzzling and at times irrational in overview. I will be expanding on the explanations at length in the general exposition.

This theory may at first glance seem superficially similar to "M" string theory. However it is radically different, and holds promise to shake modern theoretical physics to the core, because it identifies and unifies the four known forces and "more" and also explains the cosmological and universal connection between them all. It sets forth a fully universal multi-dimensionalism rather than one just for gravity. It also provides answers to most of the "hard" questions, and it provides a unifying connection from large to small by a particle theory of everything.

Why a particle theory? You may ask. There are many reasons which will be forthcoming. But the main reasons are that it reunites gravity with other fundamental forces. It explains quantum level changes via integer "steps" at not only the hadron level but at the lower fundamental levels of force particles or bosons and below.

Most importantly it demonstrates a robust mechanics of gravity and presents an extra force which has been hidden in plain sight. This new theory purports to demonstrate a better fit model of the mechanics of the universe from the micro level to the macro, and provides many answers in addressing the gravely significant "model challenging errors" being currently observed in the whole of scientifica.

The theory also explains the initial and continuing cause of gravity and dissolves the supposed dilemmas that such a weak force presents with respect to the other three (four) known forces. It provides a method of unification of all of these forces and explains how gravity has actually fluctuated throughout the life of the universe and was not even in existence for the first moments after creation.

It should not be seen as rallying an unjustified attack on general relativity. Put simply it theorizes a different mechanics which allows some elements of relativity to be utilized as it normally is in physics. However the way this theory deals with relativity in relation to space time and "time warped" frames of reference is provably justifiable and is not so kind. I must go out on a limb (and I can hear the chain saws being started from here) and state that I will be slightly undermining Newton's first law of motion because the virtual force required for a non inertial reference frame actually "does exist" by this new theory, which will be comprehensively explained.

Before you go all indignant on me and start cutting the branch, I assure you I can and will explain how we actually observe Newton to be correct at real world velocities but not at hyper-velocities. This is without resort to special relativity in order to keep everything in the inertial frame of reference. I will explain why this does not need to be the case and indeed is not.

The problem was not just with Einstein it first began with Newton who declared a law based on "mind experiment". All laws based on such experiments (including mine) should be subject to some doubt if they don't fit the observed facts. The problem was that Newton had "no facts" to observe, but in spite of such a clearly non-empirical conclusion, Einstein still formed an ironclad contract with Newton's first law, even in spite of Newton's law not fitting with the "facts" of his new theory.

Quite understandably he was faced with a dilemma (if it was as he so concluded) for an object to be prevented from reaching velocities beyond the speed of light. This would have meant the existence of a virtual force which up until then had no known cause. So it occurred to him that rather than questioning Newton, Lorentz seemed to have the answer. I.E. allow the object to obey Newton by simply changing constants such as time and distance. i.e. frames of reference.

Very few scientists questioned Einstein because they thought the thinking was "cool" and were possibly even afraid of taking on such eminent scientists, and especially Newton. Such convoluted and artistic thinking preyed upon human weakness and the mind gaming appealed to many scientists and many doubts were assuaged because of the need to display peer group intellectual prowess. P.C. or not; we currently have the end result; which is that science is going nowhere! However perhaps someone should take the bold step and question Newton and Einstein. I will do so with this theory. The only way I can do this without declaring my insufferable arrogance is to have a better fit model with which to defend my position.

The problem with the most famous element of relativity is that it firstly leaves logical dilemmas, and secondly it provides justification for the blatant and convenient ignoring of problematical facts by current science, even though they are hanging out there like dogs . . . for all to see. Believe me I have witnessed even a grade eight student asking serious questions of physicists, and astutely expressing winsome doubts at the answers she received. Students expressing thoughtful doubts such as that, are the scientists of the future and they deserve better!

Even though the narrative suggests that relativity is a "mind game" thought necessary to enable a fit of observances with the models of the day, it also explains why it can in some respects remain eminently usable, even if as shown herein to be unnecessary and even if elements related to it are sometimes not quite accurate.

With a rearrangement of some fundamentals, the model being presented solves the inherent dilemmas associated with relativity, and even explains why E only almost and sometimes equals mc^2. In so doing it develops a slightly different formula that was true at the instant of creation and is now rate dependant but universal in scope.

If it can achieve this; and at the same time offer a coherent and convincing explanation of the theoretical reasons behind the fundamentally different underpinnings of such a novel "re-think" of cosmological and quantum physics, it may well give scientists a broader yet more intricate framework in which to theorize outcomes. Perhaps this may encourage new approaches to current and future problems, with the hopeful end of the advancement of science leading to future technologies once regarded as science fiction and still thought to be impossible.

Impossible; they may very well remain, but without embracing a different understanding of the mechanics of creation, causation and the currently enjoyed but unsustainable universal equilibrium, such a negative outcome may well be assured!

This theory considers the postulation of the "force-matter" duality of a little known fundamental particle, even though already named in current quantum theories. This particle is the graviton. This theory also presents the graviton to notionally be the Higgs boson. My personal contention is that it IS, and that it is responsible for every law of gravity and motion that occurs in the universe and is a component of every particle and body in the universe. It is the unifying particle long searched for.

All I can hope to achieve with the forthcoming presentation (in such a necessarily comprehensive overview) is to arouse debate in the "academy".

If the elemental model of "earth, wind and fire" had been adhered to in the face of mounting evidential conflicts (which is exactly what is occurring with the current models), would the periodic table of elements ever have seen the light of day?

However having expressed such hope, I realistically suspect that this book will at first be met with some skepticism, most predictably (but I trust my fears will be unfounded) by the "scientist's guild" bristling with elitist indignation. Without hesitation, I will still consider such a reaction to be a positive, because nothing gives greater publicity than a good controversy which in the end is simply free advertising!

Because of the fact that I myself was "learning as I went" while writing this book, you may find a few inconsistencies between the former and latter parts of the treatise. In most cases the latter parts should have supercedence. I have taken every precaution to prevent such a thing, but you know "what" happens!

Upon reading this book you will understand the mammoth undertaking realized in the writing. In many respects it is still a work in progress and where you do find conflicting theoretical possibilities it could be because all those seemingly contradictory or different arguments or presentations may actually be reasonable theories on their own and you should reasonably evaluate them and arrive at your own conclusions.

PROLOGUE

For five or so thousand years mankind lived in awe of the heavens. He searched the sky and the stars for answers as to why he himself, as well as the stars; were even there? And he sought ardently for any profound significance. Many ideas and superstitions crept in as the ages unfolded and eventually the sum of curiosity led to a belief that the Earth was flat. Even worse; the idea that the universe moved in an unknown aspect with relation to the Earth (which was thought to be its stationary centre), prevailed.

Even after Columbus proved that the Earth was round, some "scientists" still clung to the theory that the universe revolved around the Earth.

Some five hundred or so years ago there appeared to be a degree of enlightenment and the "scientific" ideas of the time were turned on their head, subsequently it came to be determined that the Earth was indeed round and not flat, and the Earth actually revolved around the Sun.

Over the next five hundred years until now, men of thought and learning, through intense curiosity and unfathomable intellect, constructed a science of laws and formulae of unforeseeable profundity. This has led to the classical and quantum physics, chemistry and other branches of science, that have in turn led to the technologies that we enjoy today.

An important and pragmatic conclusion that we can draw from this is: Technology follows science, and whenever science loses its way or even worse, becomes bogged down in the quagmire of its own arrogance, technology will advance no further. We must resist the urge to rest on our smug laurels and become open to questioning populist theories!

I suggest that we have reached a point in history which is similar to that other "moment" five hundred years ago, which could be called a "moment of truth". This is the time when science has a choice of either stubbornly clinging to its "known" beliefs, or after deliberating long enough on the fact that things are not quite right, its scientists will be prepared to extend the limits of their imaginations to examine other models of the nature of things. Searching for one that may prove to be a better fit to the observational parameters with which they have to do, even if it means that pet assumptions are to be questioned under threat of abandonment.

As history is our guide, whenever this latter approach has occurred, (usually with traditionalists being dragged kicking and screaming to the realization), technology has taken great leaps forward and thankfully toward the betterment of mankind.

Most of us have a fascination with the stars, and we are more curious still with what lies beyond. Many people feel that our salvation either personally or corporally, lies in or beyond the stars.

It is as though we are Earth-bound misfits, in that our imaginations can extend to the realms we wish to explore but we are constrained by lack of technology to enable venturing very far at all. Even space travel within our own solar system appears to be a very daunting challenge. Conversely and paradoxically we can't seem to get far enough inside an atom and have a good look around either.

If progress in the theories we are developing continue to be thwarted by inconsistencies in the ability to marry observations with scientific hypothetical thinking, perhaps we should back pedal a little and question the assumptions we are relying upon in the development of current thought.

I will with some trepidation state; that if mankind wishes to emulate the last five hundred years and survive long enough to even have a hope of venturing to the stars and beyond, he will have to anticipate the rejection of failed intellectual endeavors in the search for correlation of force and gravity, cosmology and quantum-ology; embrace new science and take the five "hundred" year leap!

In regarding all of this, I must honestly admit that there are many things that I simply don't know; but there is a chance that if I can just help to stagger imaginations, someone else might be able to realize the impossible and comprehend a little bit further. After all, that has been the nature of genuine science, "like . . . for ever"!

I find myself still in the position where I have to ask questions such as . . . What caused time? or any other dimension for that matter? What causes the stars to shine, or mass or gravity? Are the latter really intrinsic to matter or not? Especially puzzling are the questions about fundamental particles and the behaviors involved. Some things still appear to me to be totally enigmatic and may remain so forever.

I am heartened by the fact that some stuff we get wrong, but if we learn from our mistakes and move on we can hopefully expand our knowledge of the natural universe.

The greatest enigmas of all are regarding the questions of force, gravity, mass and energy. We know what causes some forces but there are basic forces in science, the origins of which cannot be explained. I have herein attempted to explain the cause of mass, the force of gravity and the forces on objects in motion, but common to all of science I cannot explain the inexplicable and my attempts at analyzing these things as well as rethinking the relationship between matter, mass and energy may a first appear ridiculous, (or one might at least hope) novel and perhaps even intriguing in the end.

Some other questions which need answers are about the other "four" fundamental forces; what causes charge and magnetism or for that matter, strong and weak nuclear forces at the micro level? The imagination becomes almost powerless in the face of seemingly inexplicable problems such as: If an object has no mass then it should be able to be instantaneously accelerated to infinite velocity without the expenditure of any energy and therefore just by the force of "thought"! So in view of this conundrum: If a particle is believed to have no mass, is it therefore a "virtual particle? Will we ever know the answers? The trite answer to such a question would be "I doubt it" but I say to you all "Whatever you do keep searching". In the meantime I will provide some theoretical answers to these questions and many more.

Whenever a new approach to scientific theory is attempted, and especially if the title (as is the case with this book) seems to suggest that the writer might be purporting to solve all the problems of science per se; curiosity may initially be like "the bull at the gate" but the serious doubts that immediately arise will be more akin to the actions of a rancher on the other side with a "cattle prod".

The first suspicious objections would probably elicit immediate questions like: Is this writer a nut? And is this going to be just another "tick off" in a tiresome string of pseudoscientific ramblings? In a short answer I will respectfully declare an emphatic no!

It would be safe to say, that almost all physicists and chemists are aware of the disparity between the laws of classical physics and the behavior attributed to particles in some highly regarded models of quantum physics. The two sometimes seem to be totally incompatible, even to the extent that they perhaps seem to belong to two different universes.

In an attempt to marry both macro and micro science in a fairly harmonius union I have arrived at this theory, which unlike pseudoscience doesn't promote any kind of alien super science, or fanciful suppositions of the existence of parallel universes, or for that matter, otherworldly metaphysical pandemonium.

Therefore as far as the causation of origins is concerned I will not engage in any address at all, because that's another subject. The fact remains though that there are many arguments regarding origins that by my theory I have simply pushed back further in time and hyperspace and for that matter, even dimensionally sideways.

Necessity dictates though that when writing about the cosmo-universe, that both scientific, philosophical arguments and constraints must be addressed.

If you could have seen the cosmos before there was any universe at all, it may at first have appeared to be an endless void of nothingness existing in infinite space-time. It may have looked like nothing! It could've felt like nothing! And it definitely would've acted like nothing! Yet for any further sense to be made of this theory we must concur that somehow it was not at all "nothing", because it contained the potential to become something, namely the universe.

In the interest of science which attempts to deal with only observable facts and measurable parameters, I have to admit to have taken a liberty of speculation concerning the non observable realm of the cosmos. This tack is taken in an attempt to establish that the existence of the observable universe is possible only because of it, and has originated from it. If science will not allow such speculation, then I would be tempted to consider that the scientist who so objects, may actually be demonstrating the disingenuous nature of such an objection for the following reason.

Science is actually a very speculative and subsequently deductive process. It even studies observations that are determined to have occurred billions of years ago, as if they were happening today. It speculates the existence of certain micro and even virtual objects because they appear necessary to fulfill certain observational deficits. It utilizes paranormal comprehensional shape shifting to enable real time analysis of observational data in a theorized four dimensional universe.

However with the addition of the extra dimensions of this theory (which until now have been hidden in plain sight and which have come very close to having been discovered in "M" theory), this last mentioned phenomena becomes unnecessary as will be explained. Of course I am referring to special relativity.

With that in mind please allow me the right to speculate on both the macro and micro level and allow you to determine if this theoretical model is a better fit to the observances of the known universe and whether the explanations of the mechanics of the physics offered in this paper answers to more problems than other (more classical and relativistic) offerings.

Philosophically speaking; the restrained arguments are about "nothing", "something", "origins", "expanse (quantity)" and reality. Other constraints are mainly self imposed such as any arguments about the philosophical meaning of pre-existence or infinity, and the existence of other "realms" or existential universes whether parallel or not.

Speculative arguments attempting to prove religious opinions will also be restrained, because regardless of any religio-philosophical sophistry imagined by either the writer or the reader, those should be put aside "because in fact all men are in the same boat" regarding opinions of origins, in this one thing, that none of them can be proven.

There is one philosophical question which should be asked because some things stated herein might begin to appear outlandish. Perhaps we should ask ourselves. Is "what is" because of "what was"? Or is "what was" because of "what is"? If you can answer that to the satisfaction of understanding that we are only able to observe and analyze the "hows" and "whys" (no matter where they may lead) because it is an absolute and pertinent fact that without all that "was and is" we wouldn't even be here; you may then be tempted to advance further and read this treatise, even if only by way of intellectual curiosity at the very least.

The whole idea of science is not to the end of being found to be right! Einstein once stated that "The search is more important than the truth". I think that this is probably what he kept firmly in mind, because he himself new and even admitted that he didn't necessarily get it "all" right.

The cosmos and the universe by this theory are not parallel, but separate yet now intertwined. The extra dimensions theorized are declared to be parallel dimensions but only within the limits of the understanding of my "written" definition of dimensions, which will be explained in due course. It must be recognized that ALL the dimensions (of which four are already known) can be summatively "noticed" as the whole universe that we can see and scientifically analyze.

None of the dimensions are singularly visible and they don't collide with each other because they are separated in a manner similar to what we can perceive with the current scientifically recognized dimensions. Pragmatically, and without philosophical significance; ask yourself: Can you see time, or measure and observe its existence by itself? Can you see a length, breadth or height "line" by itself? No of course not. Even a cartesian plane has no thickness! While I am on the subject, I can do without mind games regarding the dimensional status of a tube or a mobius strip etc.

In considering the computer power that we have at our fingertips: I would think that now would be "the time" for a multidimensional unifying theory to be unveiled and placed "under the microscope". At no other previous time in history has man been capable of dealing with the complex algorithms that will necessarily be involved by embarkation upon such an adventure. Imagine the complexity of the subsequent "gauge" physics!

The theory I am presenting herein is a fledgling yet as broad based and sufficiently comprehensive overview that my "written word" will allow. This is all with few visual representations, and also with a paucity of formulae (which quite frankly) I will have to leave up to true physicists and mathematicians. I have tried to balance the need to make this readable and comprehensible against writing the "squillions" of pages that such a view could entail if not condensed.

Please resist the temptation to reject this theory "out of hand" as just another "mish mash" of otherworld pseudoscience. This is a serious scientific approach to the end of correlating scientific observation into a viable whole. I reject multi or parallel universalism until proofs or theories are forthcoming that have credible application and plausible models, and which can also connect both arms of physics. I hereby begin with a skeleton of an overview.

OVERVIEW

Space is not empty. It is full of matter in the form of force particles.* It is so full of them that if they were visible we would exist in a fog in which we probably couldn't see beyond a milometer or so from our eyes! These force particles are very "strange" because even though they are force, they are also matter. "Hey! Strangers are only "particles" you haven't met yet!"

*Note: In stating this it must be realized that space does not contain a medium or "aether" for the propagation of light or anything else. I will be theorizing another method of propagation.

Release of energy can cause a force against objects, which is why (if you could theoretically survive) you would still feel a shockwave in space if a nuke went off in your vicinity. You would "feel' a blast of gamma rays and other nasty "suckers," but slightly before that you might just feel a pulse of gravitons, but the impacts would both be so close together that you probably wouldn't notice the difference.* Graviton impacts would have a minimal effect unless you were very close, in which case, along with yourself, any measuring instruments would be destroyed within milliseconds *These time adjunctive affects may be measurable now-days via telemetry in order to observe whether a graviton blast would have an effect on a mass slightly before a gamma ray impact: However setting up such an experiment at the moment appears unreasonable . . . . Anyone for setting up "the atomic bomb in the back yard" experiment? Please don't do this at home. I'm a professional!

The observation that we could make here, is not that energy per se has any mass convertibility; rather it can cause force particles to move and re-enforce the release of energy at the point of impulse. i.e. such energy is released because a form of momentum force* is being exerted by seemingly massless particles which are affecting the bond forces inside other atoms at point of impact zones. This causes displacement and redistribution of binding and bonding forces which causes a forced release of BBR, convective and photonic energy. Some impulse reactions can overcome nuclear and even nucleon bonding forces. *Photons and gravitons have no mass, and the

xx | Russell Bonney

plausibility of this strange phenomenon will be demonstrated by conclusive explanations.

This activity is shown to be consistent with the pertinent laws being; firstly the law of the conservation of energy. Secondly; Newtons force law of reaction. And thirdly the universal law which states that no two objects can occupy or attempt to occupy the same space-time without the release of "energy".

Energy, motion, mass and occurrence (or not) will be shown to be the result of forces in every case.

The current scientific model of this behavior is thought to be mass energy equivalence, which supposes the conservation of energy by the "conversion" of energy to massless particles which are "blessed" with "supernatural" momentum and notional "mass" which then "converts" back to energy upon impact. (Not one religious pun intended "believe" me!)

OK we bobbed up for a bit of a laugh. I hope you took a good breath of fresh air, because we're going to be diving deep: Heads back under!

This "mass energy equivalence" thinking is allowed by a facile or "brushed off" understanding of the physics which occurs upon the occurrence of an impulse collision wherein mass should be converted to heat energy and there should be a loss of mass in the object. This has been disproved, and it takes the finite steps through the boiling and melting points of objects for any significant change to be noticed. This is a great dilemma for current theory.

This then results in having to "tweak" the relevant law to state that energy can neither be created or destroyed, it can only be converted from one form to another; "including mass" under stepped conditions. All that aside for the moment; what I want to know is where did the last bit in inverted commas originate? Because by my reading of the law, it doesn't seem to part of the proven law which only mentions energy; duh!

That's not science! I will be presenting the correct methodus operandi which will faithfully concur with the pertinent laws. This is simply an example of the academic laxity whereby science has been allowed to become guilty of contradicting proven laws of science without actually admitting it.

This cannot be empirical method; and can only lead science down the wrong path, which I'm afraid this one particular distortion of law has opened the gate to. Believe me; I am going to demonstrate that it has been a flood gate for muddy water which in one fell swoop has left scientific advancement bogged down in the quagmire of ignorance masquerading as enlightenment.

"Mass" and energy relationship will be analyzed at length but first many other things need to be addressed to enable a logical expose´ of the invalidity of the supposed equivalence, and in so doing I will be proposing the true equivalence and the mechanics behind all forces.

Considering the ongoing problems with the correlation of classical physics with quantum physics, it's probably time for a genuine non metaphysical multi-dimensionalism to "come out of the closet" because it is becoming glaringly obvious that nothing else works!

This theory is also in acceptable support of some quantum mechanics theory but not of multiverse, string or *extended super symmetry theory. I feel that it is in summation, simpler in overall context than analyzing any of its individual parts. *The mathematics involved in the predication of such theories seems to leave more problems than they solve.

Certain aspects of classical atomic theory seem to contain logical dilemmas which actually would be prevented by real world mechanical impossibility. These problems have also been "fixed" by this new theory.

Some of the "fixes" confidently undertaken in this book have no real relevance to the substantiation of my general theory of everything. They are simply necessary adjustments to keep observance in line with actual mechanical laws and to provide the theory with the required comprehensibility by consequence. If the science is not quite right, how can a person correlate anything together?

In presenting a theory such as this, if one were to in any way defy or attempt to damage any laws of physics then one would be a silly one, wouldn't one? In this theory all the laws of known physics are upheld within reason as per the context, and to the extent to which I can extend this theory, I attempt to explain how the observed and measurable effects and laws of physics are correct and actually support the theory. Moreover I also attempt to explain some of the conundrums and paradoxes apparent in physics and especially particle physics, and last but not least, how could I forget relativity?

I understand that sometimes in the real world, you simply have to ignore some contrary data or observation to be able to retain any sensible science at all. But let's please not pretend that such incongruencies don't exist. I suspect that if a better theory of stuff is not soon ensconced within the scientific forum then science may well be left "chasing its tail": So to speak.

Having said that I will now "throw the cat amongst the pigeons" by stating that I strongly contend that Einstein's theory of special relativity appears to me to be an intellectual exercise in mental gymnastics used to overcome difficulties with physics caused by various incorrect scientific assumptions. I will explain!

The theory of special relativity has been arrived at because the realities that don't require relativity for their explanation were unknown or not presented as an interrelated and comprehensive theory. The salient fact is that science required a method in which some particles and universal observations could be allowed to disobey classical laws of physics (such as occupying the same space time and support for the supposition that matter was actually made out of energy and to allow massless particles to have momentum). Analyzing and revolutionizing such concepts is an integral part of this theory, albeit not by stooping to reverse logic and applying it to a rationally dubious theory.

If you were to take the abuse that E=mc^2 does to classical science and apply it to any other formula in physics you would be thrown out of the academy on your ear. However that little detail was forgiven and the enigma was "put on the shelf" because the theory of relativity and E=mc^2 provided a method of killing two birds with one stone and this in no small way lent it serious credibility. (Oh I'm sorry; acceptability!)

Firstly it allowed scientists to explain away the just mentioned quantum level dilemmas by incorporating distorted "space time" and secondly it allowed the calculation of what they concluded to be mass by the assumption of mass energy equivalence so that an object's (supposedly intrinsic) mass was "somehow" (magically I guess) transposable with its ground state energy. The distorted space time now envisioned, led to a novel approach to the mechanics of gravity.

With a stroke of "oblique thinking" this new theory replaces mass energy equivalence with "matter/force" energy relativity and also makes a slight yet significant change to the most famous formula of all time. (Actually; the kernel of this theory began while I was sitting around a nightly camp fire for a couple of months with nothing else to occupy my time. So I guess it was really "camp fire thinking").

The theory of special relativity and E=mc^2 do however have serious logical drawbacks, which Einstein seemed to solve with his clocks and his box. (Which if you analyze the derivation of E=mc^2; it assumes light to have MASS only when moving! This is just another logical dilemma). So "all" seemed to be well for a whole century, because a "stop-gap" solution had been arrived at. In recent times however there seems to be increasing frustration in the academy, as can be deduced by the amount of money and effort being spent on the (Friggen humungous, mind blowingly massive; Large Hadron Collider!)

A soft landing on a serious note: Algebra can be used to prove any number of things if the elements and or terms of the formulas are based on incorrect assumptions when the mathematical substitutions are inserted. Any formula based on assumptions that don't fit with proven physics or don't "quite interrelate" with other classical formulae should be highly suspect!

It seems that Einstein stumbled on a formula which turned out to be essentially correct for some applications, all in spite of his erring in some assumptions.

This is probably because(as I have discovered) there is another way of arriving at a similar but expansively different formula because it turns out to really be the basic formula of the universal pattern of relationships of energy and matter over "distance", and it was "out there", and ripe for the pickings.

The only real problem that my theory anticipates at the quantum level where the magic formula seems to almost work, is that he conferred constancy on the wrong term of the formula E=mc^2. My contention is that at those levels, but at a constant temperature and gravity, E=mc^2 almost!

Whether or not this will ever be understood is of little importance to current science, but this newly envisioned and soon to be explained declaration that energy is constant and mass is able to change because of gravity differences, (with all of the implications so inherent in such an idea), leads us to a conclusion and a question or two I guess.

If I can answer the question of: What can cause mass to change? And: Why is energy a constant in a closed system? Then special relativity is DEAD! And it is then "c" that becomes a variable! The answers to all of these questions will be forthcoming and may lead science to a better understanding of the universe at both the macro and micro level.

If you are a physicist, and your scientific "PCness" is causing you to immediately discount this theory simply because it contains an attack on the theory of relativity, then perhaps you are only an intellectual and not a genuine and curious scientist. Is that a bucket I spy beside your chair?

Picking yourself up from the floor; the conclusion is that E=mc^2 is an incorrect sub form of a universal formula.

An ode to physicists the world over including Einstein who attempted right and amazed the world with his contribution to particle physics but ended up slightly wrong in a few things. Perhaps by a cruel irony he may even have set physics back a hundred years, because it remains possible for an "almost truth" to often lead astray as much as a bold faced lie.

The sheer volume and indescribable complexities of physics and other sciences which have led to the technologies we now enjoy, and with the foreseeable continuation to an even more incredible future is mind blowing. And I take my hat off to you guys. However I do believe that if Einstein had not stymied your potential cosmological scientific exploration by taking notice of Lorentz's mind games, we may already be traveling at hyper light speed to the stars, replete with gravity assisted launches and low energy requirement gravity drives. How about plasma shields and gravity deflectors? Perhaps we could also emulate "avatars" and more, with anti gravity capabilities on Earth and artificial gravities in space, and even invisibility might not be out of the question.

Please don't laugh. Perhaps you were one of the ones who took your eye off the ball. Commonly accepted theory is not necessarily correct. Perhaps you are still a flat Earther? I don't mean to be nasty but I really would like to get your serious attention!

There are enough experimental results in physics for many of you to already harbor suspicions that all is not as it seems in the universe, especially at quantum states and in astronomy, and this might leave you wondering about, not only the reasons for this but also about the possibility of intriguing future technologies should the reasons become clear.

"Assumptions are the mother of all bleepups" is a common saying. Apart from the assumptions already mentioned, here are some assumptions made by scientists which may not be correct.

A vacuum is empty:—A vacuum is not empty it still contains space which is mostly matter.

Mass is real:—Mass is caused and not intrinsic.

Energy is mass converted to an intangible substance:—Energy is just the measure of work being done or the potential of the ability for such, and not equivalent to mass! Its heat value is measured as temperature and its motional component is measured as the momentum by velocity/mass relationship. This has been found to be problematical in some areas of physics because of wrong assumptions and vagaries in the calculations of mass.

Mass and energy are interchangeable: (A distortion of the second law of thermodynamics which (by the way) is not a law in either the cosmos or a black hole.)

Gravity is an attracting force inherent to a body. This will be shown to be untrue.

$E=mc^2$: Only at zero degrees Kelvin and only if you live in a limited four dimensional universe and it only "almost" applies to a certain section of physics that refuses to obey classical scientific laws.

I am always and constantly amazed to hear astrophysicists refer to what they declare "IS" happening in the universe at the moment! Does this relativity stuff mean that you are allowed to act like a loon and get away with it? Don't worry, the public are not as dumb as you might arrogantly think they are, even as you observe them from afar safely entombed in your ivory towers!

By such behavior and also by pseudoscientific drivel being presented by respected name-brand TV and magazine icons as credible science: The whole scientific discipline is being discredited in the eye of the public every day; and you should know well the story of the boy who cried wolf! So in the same vein I will at the risk of being "un-PC" and also at the risk of being "blog bagged" by using clichéd proverbs and clichés ad hoc: I might possibly be the "unlettered boy" announcing that "the Emperor has got no clothes!"

The real $E=mc^2$ . . . ?

I have myself derived a formula similar to $E=mc^2$ by theoretically analyzing the first incrementalized instant of creation, as well as the now somewhat steady state behavior of the universe (which is what I assume might have driven Einstein at first, before he messed up somewhat, because even though Maxwell proved a similar electrostatic and magnetic propagation speed as light, (outside of false assumptions) one cannot arrive at E=mc^2 by observing a steady state universe, outside of certain contentions of this theory. Whew! What a mouthful!

The difference with my theory is that it recognizes "e" as "Ed" or divergent energy and if divergence is zero as in a perfect closed loop of energy, then mass or "c" would need to be zero which is ridiculous because it is not what we observe. So we must assume divergence to be a real number that takes into account energy losses: Then we can assert that m x Ed : "c" . . . This states that "effective mass" is proportional to the speed of light, (but only in relation to a constant energy divergence number). Conversely and within these constraints c : m, and then . . . m : "GD,"* so "c" becomes proportional to gravity/mass relationship in divergent equilibrium. *See definition.

This leads to the conclusion that E only somehow equals mc^2 on a cosmological basis, and does not directly apply to the mass—energy transformation of a less than universal size object. Paradoxically however it can be very nearly true at very, very micro units approaching zero for m and E.

Yes the energy converted by atomic fission is great indeed but not directly related to E=mc^2 . . . In fact "c^2" can be any large number worked backward that will give the required result; and then that number can be called; big constant "C" (not specific heat) which needs to have nothing to do with the speed of light. If energy mass equivalence is deemed to be correct then why not calculate the actual rest state energy mass relationship constant, and "c" can be let off the hook! The problem with this approach is that it can still only be useful at particle energy and "micro-mass" levels.

Only in a closed micro system then, can E equal m"C". (Having perhaps rattled your ire by now, I will say this in my defense: I will shortly show that "c" is not just some arbitrary speed, but that it has a very real cause which in a way is determined by the size of the universe and its energy state.

The refutation of some elements of Einsteinian physics is not a necessary part of this general theory. It simply opens up possibilities for the advancement of science to hopefully enable technological breakthroughs to be further enhanced if the theory presented herein is mostly correct.

Parts of my theory are logical assumptions which may well be disputed and refuted but please don't follow the common propensity of throwing the baby out with the bathwater. This book may be likened to a smorgasbord. Eat what you like but please don't ignore the overall spread!

Also; I will be presenting a support for the particle theory of light which is not in itself proof or disproof of the cosmological theory. I may accept a certain duality of particle/wave theory, but only as will be explained. However this "particular" presentation of photon mechanics may also help to open up possibilities for further research and new technologies.

I will also be presenting a different mechanics of electromagnetism and electro magnetic radiation (emr) which (as well as being supportive of current observances and technology), simply describes a different "methodus operandi" which does not exhibit the same difficulties noticeable in the present theories (which have been simply ignored because the "thing" works! "So leave well enough alone!") and even delving into the atomic level processes. In all of these dealings, classical laws will be unchanged.

I'm sorry! But if my theory can dissolve those troubling little details, then perhaps it may deserve a more thorough inspection! I say this with more than a modicum of respect because I recognize that I have only an inkling of the lengths that physicists and mathematicians (not forgetting other scientists), have gone to in intellectually apt attempts to make sense of the conflicts realized.

This theory provides plausible answers to a lot of questions, difficulties and paradoxes currently found in physics, even if it at first may appear a bit weird. Having taken particular care (no pun intended), I sincerely hope it doesn't create more logical dilemmas than it solves.

This theory may possibly become confused with and even referred to as a "string" theory but only insofar as the relevant bosons are being theorized to form vibrating strings instead of "round" packets of quanta e.g. photons, and particles of one dimension are allowed and prevented as the case may be, regarding existence in any other dimension. The law pertaining to this is; they must at the very least exist in three physical dimensions plus time. I also insist that forces must exist in at least two physical dimensions. Another similarity to string theory is the supposition of "lines" of various shapes and sizes across and within the universe.

In this theoretical expose there is no attempt to reconcile quantum physics and special relativity, which you know by now I deem to be a farce, or at the very least a non existent "reality?" For the sake of explanations, I state also that all particles and forces that exhibit velocity must exist in at least the four currently recognized dimensions. Single planar like "two dimensionalism" may exist in the extreme near-field within nucleons (perhaps quarks). Branes themselves have non physical dimensional characteristics in that they only exist as a point/line and only for an instant as the particle/ anti-particles digress and regress, or where particles cross to enjoin extra dimensions.

This theory can in no way explain the fundamentally inexplicable, but it may improve predictability of experimental results and scientific endeavors, and provide expanded possibilities for technological advancement.

Of course at this point I must now address the concept of the inexplicable!

The inexplicable will always remain; at least until we can get to where the forces that presume to continue the sustenance of universal existence derive from. Until then, and we arrive at the "absolute formula for everything" we haven't a clue about how the basic forces of causality and continuance of work, and they must remain outside current comprehension. This doesn't mean that we can't derive models of force mechanics and universal causation it simply means that fundamental cause is unable to be observed.

Although this is the "Holy Grail" of science, I feel that it may be forever elusive and perhaps we would be better served by concerning ourselves with building progressive scientific models of how things work, and with the fundamental overriding forces being accepted as hopefully permanent features of our universe.

The attempted fitting of models to the observations of the natural universe is a bit like trying keys in a door. You put a nice shiny key in and the tumblers don't quite line up and the door won't open.

For about a century science has been trying to turn only one key in the door which won't open. Every now and again they take the key out to give it a hopeful bit of spit and polish and in the interim someone else is allowed to try out a new key. Unfortunately we have noticed the hopeful scientists heading for the exit with a look of chagrin on their faces as they realize that the tumblers aren't falling and the door remains firmly shut.

This doesn't prevent the current batch of scientists from reinserting the now newly polished key in the door in a never-ending vain attempt to "perhaps maybe" pry the door a little.

Unfortunately the tumblers that won't cooperate are the enigmas and contradictions residing in the theoretical model that they have on their corporate key ring. They even get their keys cut at the university of their choice.

When someone comes along with the right key it doesn't mean that they are smarter than the rest. It simply means that they have picked up the right key and if upon trying that key and the tumblers fall and the door flies open, then be that as it may!

I realize that my theory will leave some tumblers jammed but the scientific method should be; to go with the theory that is the loosest in the keyhole and work on that one in deference to every other key that has been tried before.

I feel that sometime in the past a theory has been evaluated and rejected out of hand because the tumblers were a bit stiff. So with regard to this theory which builds on such prior theories: Please spray in a bit of WD40 this time, before you try the key

It matters little to us as insignificant humans, if the whole universe reverted back to cosmos tomorrow. The profundity of the power of science is not to "know it all" but to know enough to improve and continue our lot as best we can. With this in mind; let us proceed.

DEFINITIONS

Whenever a new theory is presented it may be necessary for new words and terms to be created. Also current definitions must be qualified and quantified as well, to enable rational evaluation to ensue. If you skip this chapter, you may not develop much understand of the theoretical narrative.

ANTIMATTER: Not to be mistaken for matter from another universe or realm. It is real matter but the particles within it have opposite charge signs. i.e. an anti-electron or "positron" has (according to its namesake) a positive charge. A "u" quark has a +2/3 charge while an "anti u" quark has a—2/3 charge. Antimatter and matter don't simply collide and cancel each other out in a "snuff" moment, because they are both made of real stuff and they are held apart dimensionally and protected by "agency" particles.

Because they have opposite signs they become affected by Coulombs law relationships. If and whenever they come in contact with each other they release energy by degenerating into other particles biracial particles,* which for some strange reason always seem to revert to the standard matter forms. The fact that the resultant AMOs of the universe consists of matter and not anti matter, has up until now been inexplicable. *However opposite SIGN fundamental particles may cancel out.

Logic declares that a universe made of antimatter should be able to exist. I can explain by this theory why antimatter is actually very real and paradoxically common in the universe of matter, and even though you can't see it, it has survived for a very long time in our universe (like from creation!) Note: Please bear this in mind; this is one happenstance which lends serious credence to the following theory.

ANTIPRAETOM: See praetom.

AMO: Atomic matter object, used in many references instead of objects of "mass", in order to avoid confusion.

BIRACIAL: In the context of this theory it refers to the two races of matter (true and anti). This is to avoid the longwinded stating of "matter-antimatterially".

BRANE: (membrane) the inter-dimensional event horizon at the universal and quantum level as the case may be. Objects with "atomic mass" are not able to exist in individual dimensions. They exist in the whole observable universe multi-dimensionally and so then only qualified "particle" objects existing in different dimensions are able to occupy the same space time without any dilemma, and the brane is the separation point between dimensions in space time.

COSMOS: The pre universe dimension, encapsulating and currently interwoven with the universe, both of which are being enclosed by nothing or at least anything yet knowable.

CMF: Cosmic motive force is a force across the cosmic void of the universe which is not necessarily three dimensional. (It is likely to be multi planar two dimensional). It acts across gravitines in the gravitos dimension, and is a force in this theory causing the high initial velocity of all gravitons and photons and emr. The difficulty is that I cannot theorize exactly what causes it. Similarly science can't theorize what causes "strong or weak nuclear force"! (I can't really explain it outside of the probable basic forces involved. What I really can't explain is what actually causes the force in the first place. At least I have pushed the answer to this question outside of the universe and therefore it becomes no longer anything to be concerned about at the moment).

DIMENSION: A parameter of the cosmo/universe which is either directly measurable or its effects are measurable. A dimension may or may not affect or be affected by any other particular dimension but they are all dependant on each other for their own existence within the universe. (A dimension in this theory is not another realm, religious or metaphysical idea or parallel universe).

A point has no dimensions yet for the sake of description we incrementalize it with a realizable size to explain it. Similarly a line and a plane have no real existence as they have no other dimensional measurement and they are treated similarly for the benefit of explanation.

DIVERGENCE: Stuff coming from, or going to, somewhere. A positive number is stuff going out, and a negative number is stuff going in. A result of zero means equilibrium, with just as much stuff going out as coming in. IE the divergence number changes in comparison to gain or loss of stuff. Gain is a positive number and loss is a negative one.

DRAG: Being described as an effect of certain graviton transitions, infers that the resultant velocity and momentum change is caused by loss of kinetic energy through imperfect negative elasticity collisions resulting in imperfect impulse translation of energy. In other words; gravitons pass through each other with some rebound elasticity and impulse but with a larger component of "friction". This can only occur under the dimensional law of the gravitos. It's a bit like two objects sliding past each other with friction, except gravitons pass through each other and atoms.

ENERGY: The ability to do work. At ground state or in the cosmos this is rest mass potential energy. Rest mass energy is the potential energy existing in a motionless object at zero degrees Kelvin. Potential energy is the energy of a vibrating object above ground state but not exhibiting spatial motion.

Kinetic energy is the energy inserted by force into real world objects with velocity, which is able to be expended to do work as per classical physics. Energy is a parameter of gravitons which contain force sub-particles or bosons. Energy always gets the accolades for the actions which occur in the universe. The silent and unsung hero whenever energy is released or bound is force. This little fact can lead to a misunderstanding of the processes which formed and control our world.

In the interest of understanding, the meaning of energy will be according to classical physics unless otherwise explained and where necessary.

EVENT HORIZON: The junction of different media, explained by the context. Both black holes and adjacent matter density media have event horizons. I use the term for media barriers, junctures and boundaries to hopefully lessen confusion because those latter terms contain preconceived or pre-learned content.

FORCE: the precursor of all energy expenditure. At the fundamental level I have no real idea what it is; only what it emanates from and how it is transmitted. Another thing to recognize about force is that it is permanently related to time! i.e. Instantaneous force is a logical absurdity which cannot result in any actual impetus. In the real world force operates according to the laws of classical physics.

In multidimensional theory, force is under control of the dimensions in specific combinations. The forces of this theory are; strong and weak nuclear forces, gravity, magnetic, electrostatic, binding force, and Eos repulsive force.

A particle of force can be defined as "energy matter" because the particle is translatable under controlled circumstances back to stationary rest mass energy matter. i.e. At rest state the particle can exhibit no force and therefore is unable to cause energy change. This can only occur in the cosmos.

Force can be converted to energy by causing motional changes in other particles but only in subjection to the dimension of the chronos. Without a defined time period, a force particle cannot move and convert that motion into energy by causing other matter to exhibit an occurrence of any other event, whether motional or thermal, within a specified period of time. By conclusion we can insist that force only actually operates by causing change of motion! It is the motion and transfer of force particles within other greater particles and packets of containment which cause their vibration value and force effects which are then observed in the real world.

Forces can exhibit real world, near-field, and far-field effects. Some "extreme"-near-field forces are postulated herein.

GRAVITINES: cosmic cmf lines of the gravitos. These are straight lines of the gravitos which exist across the universe. These carry gravitons in a single direction at a velocity subject to cmf (cosmic motive force) being further subject to vector force resultants of interactive transitions through other gravitons and nucleons. This causes gravitons (though not having mass) to lose velocity and kinetic energy and switch gravitines, (most probably changing direction in the same vector plane. However this is arguable for reasons to be examined).

GRAVITON: A boson theorized herein as the ubiquitous "energy" transmitting particle in the universe. A graviton is theorized to be the Higgs boson. A graviton in the gravitos has no charge or magnetic dipole or pole. (It is not a neutrino which is the size of an electron and which has up until now been thought of as an oddity, most often manufactured in the depths of stars but which plays a major part in explaining this theory. See neutrino).

In fact this theory of multi-dimensions may go a long way to making quantum physics actually not to appear to be so weird in comparison!) A graviton can also be considered to be a particle of pure "force matter" (particle boson) which is too small to be observed in any way. Unlike some other bosons it is not a sub-particle.

As a particle boson * it exhibits behavior that is never seen in any other kinds of particles. i.e. A graviton exhibits characteristics of pure force and matter at the same time. It can lose energy as a loss of velocity and matter but not size or vibration amplitude (above zero degrees k, at which temperature it will cease all motion). *It should not be seen as a particle made of pure "energy". Energy is deemed to be the ability of an object to do work as motion by force. All objects including bosons that are energetic exhibit motion, which in a stationary particle is vibration (pulsation), at a particular de Broglie wavelength which is dependant on force, density, size and velocity.

A graviton exhibits drag momentum loss, by passage through other gravitons or particles containing gravitons. It can even be stopped and eventually become totally depleted or absorbed by graviton containment objects, being specifically atoms and photons. In that case (If it was captured at maximum velocity, i.e. G-energy) it would be contained within the other particle and release exactly one "quanta" of energy. This means that a graviton "begins life" as a moving quantum with possibly a variable de Broglie wavelength of vibration as it loses energy.*

The astounding conclusion of this is that a (G) graviton could theoretically transfer the energy of Planck's constant as "friction", with the expected result of such. (It would not however be expected to be subject to Compton wavelength functions). A graviton at (G) energy and maximum velocity "y" is therefore (though subjectively) term interchangeable with a quantum. Similarly and with relationship significance, base level photons at rest will only have one quantum number of energy, however we will see that it is often given far more energy by instantaneous energy provided via the Eos dimension networking with protons to give it velocity "c". *This is not necessarily the case and remains unknown. The instantaneously acting Eos adds and removes force particles accordingly by discretional agency. Instantaneous in this case should be recognized to only be with regard to event "rise time" and cessation, but not in duration. A graviton is almost synonymous with another boson.

GD: Graviton velocity flux density: GD is an instantaneous average of an infinite number of grab samples of gravitons moving in the gravitos dimension. (i.e. within the universe).

GS: Graviton velocity deficit shadow. This exists around all AMOs.

GMF: A force causing near and far-field graviton transfer via the Eos dimension.

IMPETUS: The degree of force as a function of time. e.g. A large force acting for a short time has a large impetus. Conversely, a small force acting over the same time has low impetus. If we change either the force or the time we will see a resultant change of impetus by the formula I=f/t.

IMPULSE: Splat! Impetus resulting in an inelastic collision.

xxxiv | Russell Bonney

INTERLOCUTION: A strange effect enabled by the Eos dimension which allows atoms and particles on a near-field collision course and or also in entanglement to communicate limited data to each other. The method of communication remains unclear but is thought to be "extreme near-field force" related.

MAGNETON: A theorized single magnetic sub-particle which is unable to exist outside of other matter or the propos and which only has steady state existence in the magnos dimension. Even though there is one north magneton and one south magneton, they only attract to each other outside of the magnos dimension. Gravitons and gravitonic matter exhibit no mutual attraction or repulsion to magnetons. Magnetons within nucleons and electrons cause magnetic dipoles. How this occurs remains unclear.

MASS: The effect of force applied to some sub AMOs and all atomic and massive objects. It is hereafter referred to as effective mass or simply mass without inverted commas.

"MASS": Classical theory mass thought to be an intrinsic property with observable and calculable value existing in all matter objects which can somehow be converted directly to energy at sub atomic levels.

MATHEMATICS: The representation in numerical or algebraic forms of observable and interrelational patterns and functions in the universe. Mathematics can trick itself up and lose its way, especially when an infinite number of theoretical patterns can be theorized by changing any number of original terms. I contend that sets and matrices be left to computer and economic scientists because the real world is far too random for their realistic utilization. Further on I will show what such math can actually be applied to in this theory. Math is not necessarily transferable to observable reality.

The possible formulae for energy activity in a closed system for instance can require an infinite and complex array of mathematical starting terms or conversely just a few. Regardless of the massive and almost incomprehensible formulae which may be presented to allocate energy at any particular time or dispersion, as well as with spatial and elastic rebound considerations the end result (if the original energy component at rest was "one" then the end answer at any state is "one". What I am saying is "kiss"—Keep It Simple Sally!).

MATTER: All objective material whether directly observable or not that exists in the cosmo/universe.

MIND GAMES: Scientific mind games are recognized by the phenomena of the adjustment of constants and or unscientific distortion of effects, data and observations and the convenient disregard afforded to illogical assumptions/ conclusions term confusion and refuting evidence, in the attempt to enable a fit of current ideas to traditional science or vice versa. Such manipulation has been eminently effected by Lorentz and Einstein.

Mind games can often be noticed by the manipulation of meanings of terms whereby term confusion is utilized for agenda or outcome driven advantage. Such confusion can be noticed with the word "dimension" which is quite often confused with realm or universe as we have already seen. What is not often noticed is that it may also be confused with the term cartesian plane or surface.

I have heard of three dimensional objects such as a tube being referred to as two dimensional because it only has one external surface with length and breath. This sort of abuse of logic would then lead to the determination that a sphere is only one dimensional, and that a cube would be six dimensional. This sort of reasoning is absurd but such similar kind of "mind game" reasoning has led to relativity and E=mc^2 confusion.*

Another critically important (and I believe willful) case of term confusion is the sleight of hand involved in confusing energy with force as is done in such an example as the explanation of mass disparity in quantum mechanics to suppose that the missing mass is not just energy but some how equates to the strong nuclear binding force. Another convenient confusion is the often stated "mass" instead of matter. The term "release of energy" often includes a significant difference in meaning to the one presented in the next paragraph.

Force is not energy!. Force can cause energy to be "RELEASED" by causing work to be done. Energy is either the potential to do work by force or the energy transformation as a result of work by force. Both force and energy must originate from somewhere as individual objects that interrelate unidirectionally. i.e. Energy CANNOT cause a force. It is a "stuff substance" caused by force which is found only in matter and which forms the bonds in matter and it is the bond in the matter that contains the energy via the Eos.

A force can also cause energy transference by causing matter to move in space-time. Energy is therefore substantive, while force is not it is only an effect. Force, energy, temperature, matter and mass/gravity are interrelated in complex ways which I trust will be adequately explained in a following chapter. Mass cannot magically have a "release of energy" without a preemptive force!

For scientists to state that they can explain any force at all with current models could, if is an intentional error, be tantamount to a hoax. They also glibly state that electromagnetism is a single force. That is patently untrue because even a grade school student is taught about the two forces that combine to cause "emr". I contend that there are six fundamental forces. I.e. charge, magnetism, strong, weak, bond and gravity. Mostly however when I mention the four fundamental forces; it is in relation to quantum level physics.

*Relativity is about supposed observational disparity while the Einsteinen energy formula is not. What many fail to realize is that the formula is actually a rate formula rather than a steady state formula that somehow proves relativity. The reason that it somehow appears to work and "prove" relativity will be addressed herein.

In conclusion:

Care should be taken with mathematics, because math is not always transferable to the real world.

Unscientific mind games and reverse logic (I'm sorry if you just eyed me in the elbow!) are science fiction/fantasy and without substance when compared to empirical science and rational theories.

MOMENTUM: There are two kinds of momentum and care should be taken not to confuse them. The first one is the one you learned at school which is the continuance of motion of a body with velocity, and this then supposedly provides the body with kinetic energy.

The other form of momentum is conferred on a stationary body and more importantly on atomic and smaller particles, and it exists as vibrational momentum. This also and supposedly confers kinetic energy on the particles. This vibration is related to rest state energy according to de Broglie relationships. The effect by my theory is thought to be caused by the vibration (not spin momentum) of gravitons and summative quantities of energy "quanti". The vibrational momentum cannot be translated to linear velocity momentum on a particle with no "mass".

NEUTRINO: A fundamental sub-particle (boson) of matter which in its matter-antimatter arrangements is responsible for charge. It is also a major connecting boson similar to a gluon which is a neutral boson with no antiparticle. If the neutrino had not already been named. I would have called it a "chargon" in fact others have already referred to it. I will endeavor to show why I believe it to not be the Higgs boson or the God particle which I will reveal herein.

OBJECT: The stuff of known matter is objects described variously as particles, material, media and bodies. i.e. Every part of matter that can be conceived and therefore named: Some matter may not yet have been conceived or observed and as such is not a definable object. Conversely some supposed objects may not be matter. Not all objects have mass. Some sub atomic particles such as nucleons have "mass" others have a form of effective mass while some have no mass at all.

OBSERVER: The observer unless stated is not human, has no size and is not affected by any dimensional constraints except for the dimension of time in the sense of the chronos which is deemed to be an absolute constant. Time is constant and all else may change with relevance to time and not visa versa. If not, the nightmarish prospect of "Einsteinian mind gaming" may constrain the application of cosmo/universal physics to the limits of the inherent blindfold applied by continuing indulgence in such mental novelty. Such mind gaming has already led to conjectures that could almost be called flawless mental gymnastics. But as intellectually amazing as they are, they are still faulted by not being based on empirical formula based science.

PARTICLE STREAMS: Packetized and vibrating streams of gravitons (photons) consisting of quanti of neutrinos (neutrinos), gluons and magnetons, or quantasized "streams" of radion/magneton packets, (ramatons) neither photons or ramatons present a magnetic or electric field. Photons or gravitons by themselves cannot be affected by force fields.

PRAETOM:* A theorized pre-existing nucleon consisting of gravitons, (possibly with uud) quark arrangement, north and south magnetons and gluons providing force connections with other praetoms and antipraetoms*. An antipraetom is thought to be similar to a praetom except with a (udd) quark arrangement. Both are theorized to have only three quarks each with color symmetry and balance at rest state. (Mesons are also thought to be pre-universal). These are the precursor to neutrons at first. This subject will be analyzed in depth.

Other quarks and particles were formed from these under the changed conditions in the new universe, and many are still being formed in stars and black hole flares. * Praetoms are thought to be the precursor matter for protons and neutrons respectively.

REST STATE: state of "actual" motionlessness. See "Stationary".

SINGULARITY: Of black holes; and by this theory it is not seen to be a point of divergence to infinite density as in standard theories, rather the complete black hole contained within the event horizon.

SUB-PARTICLES: Bosons; magneton, gluon, neutrino. These are force articles which are the sub-particles of all matter including the graviton which is surprisingly of the same size.

SUBSTANCE: Any object/s of matter.

SKEW: There are two effects that can distort human observation of the universe which can therefore lead to errors in observational conclusions. These are time slew and non linearity slew.
Time slew is the observational slew caused by the delay in observational data coming from distant sources because of finite media velocity IE light. Non linearity slew is caused by observing possible non linear data from narrow time "slots" such as our few hundred year observational and measurement window here on Earth, and the thoughtless consideration of this as sufficient evidence to derive hard facts from.

STATIONARY: State of "relative" spatial motionlessness. Nothing is really stationary because our whole solar system is traveling through the universe in the region of one million miles per hour. This means that a real world object has more potential energy than a rest state object.

STUFF: Everything imaginable as well as absolutely nothing.

TINES: "Time-lines" of the photos are straight lines which carry photons at a finite yet subjectively variable velocity called "c". A photon will always emit to the tine in the direction of its formation.

UNIVERSE: The encapsulation of the visible interactive physical realm in which humans have existence.

CHAPTER 1

THE DIMENSIONS OF THIS THEORY

This includes the characteristics and laws of the dimensions ("Dimension" is not always the meaning of the word as length measurement).

In the Euclidian "three dimensional" universe, the third dimension of depth/height brings the other two physical dimensions to life. In this theory other incremental abstracts are used which by themselves are deemed not to exist. (Incrementalization and digitization are common tools in scientific description). Lines such as waves and other shapes are infinitely assumable but it must be deemed that they may not actually exist and are useful only as tools for analysis and explanation.

Theorizing of strings and extra bosons and of non existent particles such as baryons and tachyons may seem somewhat similar to theorized particles entertained by this theory. The difference is that the particles and dimensions theorized herein are deemed to not only exist but to be subject to the existing and rigid scientific laws of their own dimension. They are also able to be measured or their effects are measurable in the universe and they have no metaphysical or paranormal connotations. i.e. It is only infinitesimal size or physical characteristics and not dimensional status that may confer invisibility (non observability by any direct means) upon objects of matter existing in any dimension.

(The cosmos would almost seem to be a parallel universe; however this is not probable because its effects are seen in the universe in many dimensions and cosmic matter exists in the universe; hence I often use the combining term; the "cosmo-universe". Atoms and all universal matter are subject to laws pertaining to the cosmo-universe.)

The dimensions of this theory: (Asterisks are applied to the dimensions already declared to exist).

1

1/LENGTH*
2/BREADTH*
3/HEIGHT*
4/CHRONOS* (currently called time)
5/PHOTOS (not for the family album! Pronounced "foe-toss")
6/GRAVITOS
7/PROPOS
8/MAGNOS
9/FORCE-FIELD
10/CHARGE-FIELD
11/EOS
12/COSMOS: (Not really a dimension of the universe, but one could say of the "ultraverse").
13/ANTIMATTER DIMENSION: This dimension, if it exits would enable matter and anti matter to occupy the same space time without annihilating each other. However I theorize that antimatter when it exists does so in interdimensional brane/s.

Eleven or twelve dimensions may seem a little daunting at first but it may be comforting to know that as well as the "founding four" there are usually only one or two "in play" for any particular universal activity or causation of atomic properties and they simply allow the overlap of space time at sub atomic levels or enable actions which may seem "other-worldly" but which all translate back to the observed properties and motion of matter. We actually only notice these extra dimensions as the four we can readily observe.

THE PARAMETERS AND LAWS OF THE DIMENSIONS:

When addressing the laws of dimensions in comparison to the laws of the universe, it can be stated that the observable laws of the universe are caused by the interaction of its dimensional parts and laws. The universal laws so resulting must be equal to the sum of its parts, because such law may in fact seem to be the opposite of what may be said of a dimensional law on its own.

Quantum particles often appear to be operating under different laws than the normal laws of physics. This has often proved to be a dilemma for physicists, but with this model, that doesn't need to be the case, because the particles are simply behaving according to dimensional law, because they are actually IN another dimension/s and so subject to the law of that dimension/s.

As we observe the macro universe we are only observing behavior according to laws which are truly the sum of dimensional laws, which do not act out on their own and in so doing may sometimes seem to ignore the universal laws of classical physics.

It may still appear to some that the dimensions are parallel universes that exist with their own set of laws. This idea should be rejected, because no dimension can act alone and be observable or causative of anything outside of dimensional interactions. The dimensions must be interactive to cause the net sum of the universal laws.

The benefit of a theory like this may be in reaching a better understanding of the forces that shape our "world" and give us insights into possible limits of current thinking and enable us to find ways of *(not changing any proven universal law) but to be able to manipulate dimensional effects to temporarily interrupt certain restrictions set by the unaffected sum of the universal law. *We cannot change any laws, but what if we can prevent or enable the interaction of dimensions at will?

In such a case we in no way would be disproving laws of physics. We would be simply acknowledging that universal law is only in place because of a similarly strict set of dimensional laws. Such dimensions are also affected by forces which don't necessarily appear to directly apply to the universe itself. Again universal law is seen to be "resultant" law by the summative interaction of law sets.

The salient idea is that if we are legally prevented from changing any of the laws themselves we may be able to manipulate the sets in a way that we might be able to vary the sum of law sets acting on particular quantum particles at will. In this way we may be able to temporarily adjust or interrupt the behavior of universal objects and perhaps in the future even on the macro scale.

THE OVERSHADOWING LAWS OF THE UNIVERSE:

The first law states: Nothing can occur without causality and everything that occurs is by the action of a force/s regardless of the point of origin. The second law of the universe is: Any effect that causes consistent observations which result in a law being evinced is not itself necessarily counter-subject to the said law or effect.

The laws of the cosmos: The cosmos exists only in the physical dimensions. Within those dimensions exist praetoms (from pre-atoms). Praetoms have potential (as rest state?) energy of some degree above that of an atom. The relative amount is to be presented in the theory. The cosmos is not in the dimension of time and is at zero degrees k and therefore is not subject to the laws of thermodynamics. There is no motion within the cosmos; however the cosmos can be seen by our observer. It only exists because of the physical dimensions in which it is contained. The laws and possible options of the cosmos can be likened to a sleeping giant, laying dormant until wakened by a threat to its existence.

The first law of the cosmos is. Should it become damaged it will attempt to repair itself instantly. The second law of the cosmos is it will exert an instantaneous and continuous cosmic-"velocity" motive-force (cmf) across a rent in the attempt to instantaneously close it. (Instantaneous motion is not measurable as motion because it takes no time and therefore also expends no energy). So the third law of the cosmos is: Motion is not possible in a steady state cosmos, and energy is unable to be expended, such motion would be concluded (without further explanation) to be self contradictory.

LENGTH: Length is a straight line of infinitely small width and height stretching across the cosmo/universe that cannot be bent.

BREADTH: Breadth is a straight line with similar characteristics stretching across the cosmo/universe at right angles to the length.

Together they make Cartesian planes and sets of planar lines of which there is an infinite number.

HEIGHT: Height or depth is also a line that cannot be bent bringing dimensional matter into existence only by enabling objects to be identified having been formed and currently existent in three dimensions. The cosmos and universe themselves are such objects.

These three dimensions are called the physical dimensions and they are interchangeable with each other and are also incorrectly referred to as the dimensions of the universe or space.

CHRONOS: Time in its elemental and absolute constancy. The first law of the chronos: Time allows motion and work to be done by energy and force. (No time no motion) The second law of time is that no objects in the same dimension can occupy or attempt to occupy the same space at the same time without the release of energy or work being done. The third law of the chronos is: Objects colliding at velocity will be subject to vector law of the force because the force-field dimension is existent within the dimension of time and all collisions between AMOs are in someway inelastic, so their motion within time will change. The fourth law of the chronos is that time is a constant regardless of frames of reference which are only mind games.

I must address the other forms of time that can confuse the issue of understanding the chronos.

Observed time: Related to the human observance of time from life to death. All other mental determinations of time are relevant to individual experiences and frames of reference. Observed time is not an absolute constant and lends itself to mind game confusion. Such things that have been stated are: "This only appears to be the real time to you and your measuring devices with which you measure the universe. To someone else in a different location: His observed time and measurements may be different yet real to him". This sort of thing does no lend itself to scientific understanding of the universe. If the whole observation and measurement deal is that subjective then we might as well give up now and go fishing! In the case jut mentioned: The time doesn't change, it's just that different parameters at different locations are what may cause an observational "rate" change.

Measured time: This time really is an arbitrary number based on observances of our own planetary motions. (On mars the second, day and year are different). However it would be a mistake to assume that because the Martian second is different, that this would disprove $E=mc^2$ on Mars. This is because the relationships within the calculations of scientific formulae using the Martian second substituted in them all; would end up with the same result and still remain an enigma for the "Martian Einstein" to evaluate, and the same disputable theory would be the result.

The measurement of planetary time has been superceded by cesium clocks, which is a patently obvious irony when you consider the laxity with which relativistic science treats chronos time. The other thing about atomic clocks is that they are relative to gravity and temperature. So it adds insult to injury to suppose that measured time should be treated as a meticulous constant. The only constant is the chronos.

This is not a mind game; all things being equal, current generation clocks are phenomenally accurate and therefore have a close relationship to chronos time. In this theory the chronos is deemed to be constant so an understanding about the real variables can take place. The reason that cesium clocks must be reset for space travel, specifically GPS satellites is not because of relativity or the lack of gravity they are experiencing but because of the change in GS.

Measured, observed and chronos time have all mistakenly been taken as extra variables when attempting to understand the so called space time continuum.

The Einstein—Podolsky—Rosen criterion has also been used in an attempt to understand certain enigmas with physics and the apparent disobedience of quantum physics to the standard laws. My theory will readily address this problem. I have no problem with Einstein's gravitational theory (in that it provides a model which generally seems to fit with observed results). or with is slit per se, only with his box and clocks. My theory will show that quantum physics does not need to obey the laws of physics because it observes particles, some of which are the multidimensional CAUSE of the laws, and causative functions are not subject to the laws they cause as per the second law of the universe! This then should render special relativity to be treated as obsolete science.

PHOTOS: The "foe-toss" consists of lines of containment of photons which are ruler straight and without any other laws of dimensions interacting with them the photons will remain on their line within the four classical dimensions. The speed of light is determined by emitting protons which by interlocution with other protons in closed density proximity determine the cosmic temperature of the universe by GD and Eos parameters and emit light at a speed in proportion to the Eos data value of GD in the continuing attempt to enable the most efficient return of atoms to the cosmic state as praetoms.

GRAVITOS: The gravitos exists in and is subject to the physical dimensions. It is also subject to the chronos and force-field dimensions but none of the others. It is an enabler of the propos and interacts with the photos to effect gravity and mass as will be explained.

PROPOS: The propos (for propagation) is controlled at the quantum level by protons and most probably other particles. The proton determines the speed of propagation of charge, magnetism, "emr" and light in its vicinity by taking the cosmic temperature via gravity. (To be explained). The propos is the medium of "field wave" propagation. (I have coined this term to replace "emr" in most instances to avoid confusion, because they have conflicting definitions).

This theory theorizes that the upper spectrum of the "emr" continuum consists of photon particles, which to some extents is reasonably explainable by wave propagation theory which propagate via the photos and "emr" which propagates via the propos. With such separation it can be concluded that the two won't interact with or affect each other.

Wave propagation may indeed exist as a result of forces acting on fields projected into the magnos and charge-field dimensions, from zero hertz being a steady unilateral force up to an unknown maximum frequency but these are subject to permitivity and permeability laws.

This inter-dimensional mental conundrum is caused because of the difficult notion of multilateral dimension effects. Such an idea is best explained by the old story of the flying astronaut in the third dimension, traveling above a single plane universe. He does not affect the flat universe and it does no affect him; but by its existence it allows his flight in three dimensions. Imagine if he touches down on the flat universe. What would the flat "universians" observe? Could he affect them? Could they affect him?

Atoms above zero k are in constant motion. Photons as individuals or in strings, both polarized and random, and atoms are affected by bi lateral force field effects. (This brings us to the dimension of forces and fields). The interlocutions between atoms and molecules (with or without crystalline structure) results in some bizarre effects on light, especially at media event horizons.

The interlocutions between protons themselves and photons (approaching or departing) create vector force fields that can become patterned, oscillatory or diffraction anomalous. This is not surprising if elasticity and nodal effects (which are common in physics) being caused by accelerative, deccelerative and elastic multidirectional forces are operational on any object.

However chaotic things might seem; randomness is never an option under these circumstances, so such things as double diffraction, the Goos Hansen effect, spatial displacement and other effects are not surprising and are presently giving rise to technologies. What I find as surprising is that with discovery of the "optical tweezers" effect, thoughts of anti gravity don't appear to have even arisen!

At this point it is necessary to discus protons in particular. The protons of elemental atoms are theorized to all be dimensionally different (because of sub-nucleonic and perhaps other affects) even though they might seem superficially similar. They can be thought to partially exist in different dimensions, and all elemental nuclei will therefore have differing vibrations and other characteristics, and their effects on electrons and photons may be vastly different, dependant on the quantum dimensional parameters, as well as the universal state, temperature, nuclear and dimensional states they exist in. How does light really pass through solid matter?

MAGNOS: This dimension and the charge-field are the only dimensions that exist only in proximity to atoms and AMOs, being proportional to the atomic density of the atomic objects. All Atoms exhibit forces and fields which include both electric and magnetic. An electric field is a static field that has a "build" velocity (per Maxwell) and depletion by inverse square law in proportion to distance. It is an open loop field of radiating lines of "energy" that exhibits an electrostatic effect which can be directly measured.

A magnetic field is a "force" field with similar build velocity and attenuation as an electric field. It is usually somewhat angularly vectored opposite to the electric field and it cannot be broken to facilitate the insertion of measuring devices. It can only be observed by effect. These fields are both mutually exclusive but co-effective. The motion of electrons is causative of magnetic fields and magnetic fields are causative of the motion of electrons in a conductor, but NOT in space. However the electric field is caused by external electrical energy input to a conductor and both fields are energized by protons and to lesser extent neutrons.

The magnos contains an almost infinite number of field-lines existing in four dimensions of space-time. These are the basic framework for the existence of magnetic fields. Magnetic fields differ from electric fields with regard to the first fact that the size of an electric field being governed by inverse square law over distance is determined by the value of electric field-strength. This is regardless of the size of the charged object.

Conversely and significantly for this theory: The size of a magnetic field within the confines of a field decay power law is determined not by the strength value of the source of the field but by the size of the object exhibiting the magnetic dipole effect.

The magnos dimension and the charge-field are both dependant on the force-field dimension but not visa versa!

FORCE-FIELD dimension: The force-field dimension is responsible for the dissemination of gravitons (energy) between nucleons, and from nucleons to electrons. This energy transfer occurs at a rate determined by this dimension and is then called convection and quantum number respectively.

The convection does not appear to occur because of an atomic density-proportionality relationship. I contend that its determination is more dependent on the time taken at "c" for each atom to complete its own charge and magnetic field transfers to electrons. The time taken to do this is expected to vary from atom to atom and affects the coefficient of thermal conductivity (specific heat) of various materials. (This idea may need to attract further research). The force-field dimension along with the Eos also underpins the forces that include weak nucleonic and strong atomic binding forces. It also controls the transfer of vibrational energy through atoms in a manner which obeys vector law. It is also the dimension through which charges and magnetic fields propagate at "c".

CHARGE-FIELD: The dimension responsible for the application of charge to particles. Charge particles of two different types are intrinsic to ALL greater particles. The resultant measurement of charge on a particle is due to a surplus of one or more—ve or +ve particles. Particles within the dimension are massless and they have no attraction or repulsion to each other in a motionless state or in a fermion or packet. It is only when combined with the force-field dimensional characteristics of energetic particles that this can occur.

In this case the charge-field and force-field are dependent on temperature (The measurement of the energy state) such that the net charge in particles is temperature dependent to some degree. It is unclear whether this can be said about the relationship between the magnos and the force-field with a similar resultant temperature affect on magnetic fields. This perhaps should not be confused with a magnet becoming demagnetized by heat, although the connection might be found to be significant.

This total force-field dimensional effect is that which allows electrons to lose charge and magnetic particles in a synchrotron and emit at a tangent from a circular path as electron-photons. Photons and I suggest to some extent electrons emit to tines which are straight and tangential because of conservation of energy and not angular momentum law as you might expect.

This is likely due to the magnetic field being sufficiently strong to cause the alignment of the magnetic dipoles and splitting them apart. This would immediately release energy and heat the affected electrons to incredible relative temperatures such that they could be thought to lose their charge particles entirely. This suggests (which is explained in depth further on) that electrons as well as a magnetic dipole, also consists of a packet of gravitons which is (by trans-dimensional definition) a photon.

The photon so emitted to a tangential tine is of very high energy and vibration frequency. In such a case the electron-photon beam can be extremely powerful and polarized. The polarization occurs because the dipoles are stripped in the same direction as the causative magnetic field and the vibrations produced by the charge particle emission are therefore corporately planar.

EOS: This dimension is not to be confused with the charge field which operates with the force-field dimension. The Eos is not affected by time. It provides the atomic nucleus charge balance across the universe and universal "GD constant" data to the same. It is also the dimension which at first caused matter and not antimatter to be the atomic nature of the universe, by the reason that it controls the charge in quarks. The cosmos was the complete and somewhat incomprehensible object of cosmic matter, and the Eos is an inseparable dimension of the cosmos.

Local emfs and charges can temporarily override the Eos because the force-field dimension and the magnos take precedence within bonded atomic matter. If the Eos didn't exist it would be likely that spacecraft going to other planets could be struck by lightening because of the likelihood of charge differences between planets.

A supportive reason for the postulation of the existence of the Eos is that the whole universe appears to be "Earthed" or "grounded"; (to put it in electrical jargon).

The reason it is determined to be instantaneous is that there has never been any noticeable short term mass, gravitational or any other anomalous effect or any other dimensional characteristic (that cannot be explained by standard physics) which has been measurably observed upon the cessation of any emf or force, which can demonstrate any remaining energy or binding force deficit in any nuclei. (Not including ionization which is quite different). Note: Charge transference between atoms in near-field relationships is under the control of the force-field dimension and has been shown to occur at "c".

Some weird effects discovered in quantum physics can be explained by the Eos. A law of physics states that no two adjacent similar fermions can be in the same state. i.e. *by way of simplification for this example; if one is up then the other must be down. Now all loose particles and atoms in the universe are connected together by the Eos. So two adjacent fermions have a direct connection; if they exhibit this inverse bilateral relationship and you move them apart, and then change the state of one individual fermion, the other one will flip or flop in inverse concert. This is because one; the Eos connection is instantaneous and two; because of this fact the two particles are never actually apart until they become absorbed as part of a more massive body. *There's plenty of time for the quantum twins Pauli and Fermi later!

Some analysts of this effect have concluded that no data could have been exchanged, so they must then have a metaphysical or religious reason for such occurrence. Of course we can conclude that data was indeed transferred instantaneously because in the instantaneous dimension of the Eos, the particles are never "actually" apart regardless of any observed distance. Connection with "near" electromagnetic fields is perhaps the only thing that can release them from the Eos, and in that case they can then become normal again. Whew!

With sub atomic particles this effect is called synchronicity, entanglement, or non local correlation. Recent studies have shown the effect to occur with two atoms as well. In this instance the effect was termed interlocution.

Accumulation of near-field atomic and molecular charges due to electron depletion or overpopulation even to the resulting far-field effect of lightening are examples perfectly explainable by normal physics, (but with a novel explanation to be presented herein), and are caused by charge field effects.

The proton is able to combine the fields in spatial patterns at various frequencies dependant on internal and external forces and fields to facilitate propagation of the field data via the propos. The electron is thought to be involved in the creation of a field/s but it is deemed to play no part in any emission of either a wave or particle for propagation whatsoever.

The Eos enables oscillating electric and magnetic substances to have their energy passed into the universe at the speed of light subject to pertinent laws. Apart from the formula $E=hf$, I am making some changes not to Lorenz or Maxwell laws etc, but to the method and mechanics of transmission of electromagnetic energy.

I do not know why the propos has the same propagation speed as the photos except that I surmise that they are both caused by the same reason (which will be addressed below) and unlike Maxwell and Einstein, I do not conclude that light is an emr simply because it has the same velocity as emr.

This theory dictates that the propos is a different dimension than the photos, and I must admit that at the quantum level the dimensions become almost like parallel universes. The Eos is theorized to provide the actual force affecting the photos and causing the velocity of photons and therefore giving them the apparent kinetic energy of momentum. In a following chapter I will present an expansive description of the operation of the Eos.

Here come the "big spanners in the works" which I have thrown in by this theory!

1/ Mass: Mass is not real. Matter is real. Mass is an effect.

2/ The laws of motion are caused by the same force that causes mass to have effect.

3/ Hadrons, photons and lesser particles have no real mass because they are partly causative of mass and for the reasons.

As you are already aware neutrons are not a useless appendages hanging out with a protons. They are a main receiver of mass affect. Inferred mass is declared to not be proportional to the graviton density of nucleons but dependent on the numerical number of nucleons per cm^3 and inversely proportional to the density. This also infers that inner nucleons of any object will have slightly less mass and a slightly lower binding energy.

Also by the same principle observation, larger atoms are declared to (proportionally) have slightly less overall binding energy per effective mass than smaller ones. This effect can be noticed by a cursory glance at the standard binding energy curve.

This reduction effect only begins to become noticeable when the nucleus is large enough to have the inner nucleons affected by the GS of the outer nucleons. (Note: This will not substantially affect the "mole" because that is a comparison of grams of substances, and both substances will be affected fairly equally within density proportionality relationship).

Some deviation from the curve will be readily noticed, especially with smaller atoms and with larger atoms for differing reasons which should by now be obvious, suffice it to say it is determined to be dependant on firstly; "fewer sample sets of nucleons" in smaller atoms, and secondly; the "quantity of neutrons and even their position in the nucleus" for larger atoms. Mass differences will also be noticed with the various isotopes of course.

The H1 atom of hydrogen has much lower (unused) binding energy and lower "mass" because it only has a proton in the nucleus and so it only has proton plus electron "mass".

However it can be shown by "band spectra analysis" that the neutron has more "mass" than it should if you were to simply add an electron mass to a proton. Two specific examples measured in both energy units and mass units are as follows.

Example one: In energy units (using $E = mc^2$), the masses are: Proton: 938.272 MeV, neutron:

939.566 MeV, mass difference = 1.293 MeV, electron: 0.511 MeV.

This is not binding energy; it is ground state "effective mass" difference, with binding energy left intact. The binding energy is significantly greater at about 28.3MeV

Example two: in kgs. Mass of hydrogen nucleus (proton plus electron): 1.672649e-27, Mass of deuterium nucleus: 3.343637e-27, which is 0.001661e-27 heavier than by doubling the mass of the hydrogen proton/electron. Again this cannot be binding energy.

Now these effective mass differences will not affect binding energy by very much; it's just the fact that they are even observably different which lends support to this theory, (as does the binding energy v nucleon density curve).

Of course binding energy is real*. This theory understands that it is the energy which will be required to force the nucleons apart. However this theory sees energy as caused by gravitons which do not have mass, (even though they are particles of force). * Also the theory is not attempting to refute nuclear physics. It is simply about the mechanics of tying together mass, energy, gravity, motion, magnetism, electricity and a few other things.

When the nucleons return to ground state their apparent loss of mass is directly caused by graviton loss. It's just that now they are left with graviton density deficit, which in calculations of "mass v energy", the formulae and ensuing results will be similar. However the mechanics are completely different.

In the case of a proton existing in the gravitos dimension alongside the neutron; the proton must transfer gravitons (basic units of energy) to the neutron for parity reasons and even reemission as BBR unless it instantly receives sufficient energy "quanta" to emit a photon or more. (Especially to meet requirements of the "Pauli exclusion principle", this even applies to hadrons). This "inter quanta" BBR emission may also be the reason for "mass" anomalies in the periodic table and with certain molecules.

This means that missing effective mass is not necessarily related to missing energy it's just that the mass quantum levels are not yet realized. I find this as particularly strange, for a science which theorizes energy "mass" equivalence to recognize energy quanti but not mass quanti. This is evidence of an element of tardiness in the understanding of the fundamentals of that particular theory.

The effects, laws and observances explained by this new theory involve the particle theory of light and mass/gravity propagation. Below I have presented some of the science being explained by my theory:

Why stars started and continue to "shine"; drag in space; red shift in outer galaxies; why the outer galaxies appear to be accelerating outward; why the night sky is not a sheet of background light punctuated by stars; what causes mass, gravity and black holes; what causes the laws of motion; the anomalous behavior of binary pulsars and why most galaxies have the shapes they do including the enigmatic spiral form; Why large planets do not exist in the inner solar system; why the Moon appears to be moving away; plasmas both cold and hot; the behavior of light including lasers, diffraction, absorption, opacity and translucency, light nodal observations and interferometry including Einstein's slit. A particle theory of gravity and its speed; The creation of gravity by light; the effect of plasmas on gravity and light; quantum physics anomalies; what causes "quanta" and why quantum integer steps apply to everything except sub-fundamental particles; what happens to photons traveling in a string or packet when entering a media event horizon; the real reasons that cesium clocks must be readjusted to enable GPS satellites to function accurately; why planets seem to create their own heat; why the kg standard in Paris has lost mass/weight. Why the Sun is cooling down and; what are radio waves, really! Why the theory of relativity is a "croc" and much, much more.

When a new theory is presented that attempts to provide answers to current questions it may elicit more questions than it answers. If this is the case then the theory should be immediately suspect, especially if the original questions remain unanswered. I must admit that this theory does rouse some further questions but I propose that it answers many more questions than it creates. Some of the questions that remain are traditionally obtuse.

Some such questions are. What causes the protonic photaic force if photons have no charge or magnetic field? Ok then: What actually is charge and magnetism and what causes them? What causes any force at all for that matter? What prevents electrons from simply crashing into the nucleus because of Coulombs law of attraction? How can electrons take it upon themselves to drop an energy level and emit a photon?

I would be treading on dangerous metaphysical ground and perhaps even find it necessary to use the "G" word if I were to suggest that there must be pre-eminent and proactively invisible yet unknowable reason for causality. Notwithstanding this, the questions remain open.

CHAPTER 2

THE BEGINNINGS OF THE UNIVERSE

This is a speculative theory of origins: The simplicity of this speculation is for the purposes of overview only. To attempt a plausible fit into models of quantum theory seems as daunting as quantum physics itself. Many intricate relationships and forces of various strengths retaining and releasing particles of various integer states as theorized in quantum physics, became infinitely varied as the universe cooled. I believe that uncertainty principle is applicable not just to velocity and spatial position but also to extreme temperature states with resulting effect on quantum states and even Pauli Exclusion Principle and Fermi state/level effects.

This speculation is not about origins of everything. Whether or not you believe in creation or self emergence of stuff is beside the point. I will admit that logic causes me to accept creation as the cause of stuff. Any other opinion remains valid if logic is seen to be the crutch of the weak minded.

Metaphysics and fantasy are the stuff of the mind and can lead to scientific discovery only if the theories so concocted can lead in a mathematical, scientifically and empirically legal sense to acceptable fact. Whether this speculation concurs with these criteria is for you to judge.

The following is deemed to have occurred instantaneously at first and then began to rapidly slow but not necessarily at a determinable or steady rate. I must therefore arbitrarily incrementalize it for at least the time it takes to explain it.

The cosmos existed in a timeless cocoon of "three dimensional" space of unknown measurements. It existed at zero degree k without motion or energy expenditure. It was a cosmic sized black hole existing in the hyperverse. Even though cold dark and motionless, it was however subject to laws which it had never before been required to use because nothing had ever moved or changed as far as we can tell.

In an instant "something" happened that changed the cosmos forever. The cosmos was shattered by a violent force and an immediate discovery of another dimension occurred. The other dimension was the Chronos. It appeared as an anomaly in the cosmos which had never before experienced motion. The cosmos now discovered the laws pertaining to "itself". The first law it discovered was the law of self conservation.

So the praetoms of the cosmos attempted to instantaneously close the rent, and upon moving over any distance they found themselves unable to move at infinite speed because of the Chronos and it was then they then discovered other laws.

Motion requires the release or use of energy and by moving at a finite speed they also found realization of the existence other dimensions which also came into affect as a direct result of the requirement of the praetoms of the cosmos to obey the laws of their own dimension.

Another law soon discovered (when they moved into the "void" in their dimension) was that when some praetoms attempted to occupy "the same place at the same time" they found that they lost energy because of drag and they also decelerated and experienced directional change.

This all occurred during the first moments of creation whereby the cosmos discovered time and motion and kinetic energy. The praetoms having lost energy could no longer remain praetoms and they became something else called deuterons. The surplus matter they contained was instantly turned into energy of all descriptions which headed back to the cosmos at almost infinite speed.

The energy released was phenomenal and to an observer a blinding flash of universally wide light would have been seen. This process and the derived light emissions occurred at countless points in the universe. I suggest it was not a "big bang" or an explosion of matter and energy from a single point; it was a vast shattering caused by some sort of semi-elastic collision of indescribable force!

So to reiterate more concisely: Instantly upon losing some of their potential energy, praetoms ceased to be, and became at first deuterons, single protons or shattered neutrons at very high temperatures and then as they lost more energy they became atoms many of which formed atomic bonds and molecules by now known process and the utilization of many different quantum particles of cosmic stuff. Quark gluon plasma would have been quite extensive at the time and as it cooled would be able to provide many of the "building blocks" for the new universe.

These new atoms now had less potential energy. Here I will attempt the difficulty with the speculation as to how exactly praetoms became atoms.*

The best way to address the problem is probably to consider the laws of dimensions. We now find that the cosmos is being affected by the dimension of time which states that objects cannot occupy the same space at the same time or else . . . Praetoms found themselves in this position and in so doing lost energy. A post cosmos praetom is deemed to consist of the matter and energy of one neutron and two protons. A deuteron consists of one neutron and one proton. A H2 hydrogen atom consists of one neutron and one electron and one proton less the mass of the electron. So the praetom lost at least the equivalent energy of a whole neutron which is about twice the theorized energy of a cosmic photon. In other words an enormous amount of energy at the quantum level. *This process is treated in greater detail in chapter 16.

Nucleons which were of two types; an "udd" quark arrangement and an "uud", began to decay. Neutrons decayed rapidly but protons found that when they had decayed to the point of losing an electron any further decay was truncated. And whether or not they were being forced or bound together with neutrons their magnetic dipole and electric field caused a natural interaction with the electron which had just been emitted, and an atom was formed.

Considering the laws of randomness; the likelihood of praetoms becoming atoms of low complexity such as hydrogen and helium was high in comparison to their chances of becoming atoms of high complexity and this is realized in the currently recognized spread of atoms in the universe. The nodal effects caused by the deccelerative effects of drag and the interference nodes caused by convoluted randomness resulted in the creation of clouds of lesser and greater atoms giving rise to the periodic table of atomic "densities".

This allowed the collection of massive clouds of hydrogen and or helium and lesser clouds of denser atoms that (because of nodal constraints) found themselves gathered into spatially positioned groups by a process called accretion. Accretion is thought to occur by electrostatic charges being formed when particles rub against each other. I consider this approach to not be well thought out and I contend that it was simply by magnetic and charge field interactions. Sub atomic fermions have dipoles, and most matter exhibits some form of magnetic and electrostatic charge field. This is not readily observable on Earth because of the stronger force of Earth's gravity.

At this stage of the "instant" there was no gravity or mass and the temperature of the atoms so formed after the first massive emission of energy was most likely close to the Bose Einstein condensate so they didn't fly apart. (The reason for this should become apparent by the theory even though standard physics could not allow this to occur. Remember this is a non standard moment. The universe is in the first milliseconds of creation!) *

The new atoms attempted to follow the law of the cosmos which bound them to return to the cosmic state of motionlessness at zero degrees Kelvin with rest energy once again equaling potential energy. After the initial release of energy the atoms attempted to continue sending themselves back to the cosmos by radiating a much BBR as possible as well as by packaging themselves up into "quanta" of matter consisting of the smallest sub particles of praetoms which I suspect are gravitons* (which are also theorized to be the singular particles of force and "sub quanta" BBR). These parcels of energy "quanta" have already been noticed and have come to be called photons. *Some gravitons and or their sub-particles () are emitted by black body radiation BBR, but the majority by far (at elevated temperatures), is caused by a process which will be explained shortly.

Other means available to the atoms were undertaken in a futile attempt to return themselves to cosmos by the utilization of other propagatory energy in the form of electro magnetic radiation propagated via the dimension of the propos, with BBR being emitted via the gravitos or Eos depending on the particle being emitted. (This assumes that neutrons may actually do something after all: i.e. radiate gravitons and bosons as BBR).

The dilemma atoms have in returning to the cosmos is that they have two ways of attempting this which are self defeating because of their mutual exclusivity. 1/ They can attempt to return to zero k by sending themselves to the cosmos by photonic and propagative energy. As they have no intelligence they do not realize the law of gradually decreasing ability with loss of energy and have found themselves under the third law of thermodynamics and if left alone to their attempts they will eventually find themselves almost totally devoid of the energy requirement to become Praetoms again. They could then find themselves almost motionless and existing at a temperature somewhere close to the Bose Einstein condensate temperature with infinitely unavailable energy requirements to complete the journey. (Unable to emit more energy they will be lost in cold dark space for an infinite time).

Conversely and paradoxically atoms are able in some circumstances to absorb enough energy until they reach the temperature realized at the event horizon of a black hole (which I will humbly call the Bonney significance temperature "BST"). and suddenly find themselves with the full amount of energy required for them to return to being potential (sleep mass?) energy filled praetoms again and become assimilated back into the cosmos. Which this theory suggests is exactly what is occurring in black holes and also at the outer universe/cosmos event horizon.

The Universe is surrounded by the cosmos or hyperverse but because of the dimensional interweave of the cosmos linked gravitos with other dimensions in space time the cosmos is also inside the physical dimensions of the universe in the form of black holes which are dimensionally connected to the cosmos/hyperverse. So if a person could enter a black hole or the cosmos without damage and with the right cosmological protections and cosmological velocity and navigability, one could reappear at any other point in three dimensional space via another black hole in an instant. This appears to be currently out of the question however and does not in any way allow the idea of time or warp travel.

(The following at first glance may sound like psuedo-scientific "clap-trap" but if you are a nuclear physicist please think for a second). This proposes a theory of a closed loop temperature range which makes just above zero k the same tipping point as the extremely high temperature realized at the event horizon of a black hole. This is not so unreasonable when you understand that temperature is not "brrrr or whew!" Temperature is actually the measurement of the vibrational energy of a particle which is actually being caused by the jostling of a variable quantity of force particles. (In this case nucleons about to enter the cosmos whereby they will instantly cease all motion and in so doing lose all temperature.

Putting it another way: This is only a paradox to the human mind and not to an atom attempting to return to equilibrium by either a conveniently positioned black hole and the subsequent realization of maximum energy, or the unbeknownst futile attempt to return to zero k. (As soon as an atom converts to a praetom in returning to cosmos it instantly reverts from the BST to zero k). This seems to be impossible but not if the kinetic energy of vibrational momentum becomes instantly converted to potential energy by the atom's ceasing all motion and also losing all physical "mass" by reverting to praetom matter existing once again in a timeless state. (I will not even attempt to theorize the exact mechanism of praetom reunification from atomic parts). Note: To avoid a possible logical contradiction which may be noticed, I will allow that the whole cosmos was fully engaged in becoming the universe and that "pure" cosmos may never exist again. This will allow for the lowest temperatures to be expected to exist in both black holes and empty space, to be a few degrees above zero k, forever!

Fortunately for our own existence, atoms being the dumb "animals" that they are, have undertaken to attempt the former course because the only recourse to an atom without sufficient "positive divergence" external energy input is to attempt to shed energy any way possible. This leads to the fact that atoms give each other energy and attempt to at the very least, to arrive at a state of relative equilibrium with their surrounds. Far from being "kind and sharing" they are attempting to rid themselves of this horrible thing called temperature being any degree at all above zero k. (I admit to a necessary digression and this subject will be treated at length in a later chapter).

The attempt of atoms to return to the cosmos via a continuous stream of energy over the tines resulted in another "unforeseen" effect which caused the whole process to slow, which also caused the progress of the universe towards the state that we humans exist in.

What happened is that the massive streams of photons collided with each other, and although photons have no mass because they are causative of mass, (the theorized reason will be soon tabled) they are still matter and subject to the first law of the chronos.

When they collide, (especially head on) they lose some energy and significant velocity and incur a non specific directional change. How this is theorized to occur is that they instantaneously switch dimensions in avoidance. The photons instantly and without spatial motion cross the brane and move into the gravitos.* In so doing they emit a particle called a graviton. The graviton is subject to the laws of the gravitos and is emitted at almost infinite speed. (Some scientists have speculated ten to the tenth times C in certain circumstances). * This will be a nice headache for eigenspace dimensional calculations).

With the massive and almost infinite quantity of photons colliding at almost infinite speeds a vast quantity of gravitons was emitted. These gravitons according to the laws of the gravitos and time found that when they collided they lost energy and velocity through drag on each other. By drag they were perceived to have a vector force change in direction.

So now you have re-emitted gravitons which were at first heading for the cosmos in all directions but some are now being forced to move around in universal space and slow down. The amount of slowing can be surmised by the amount of time it takes solar gravitational anomalous effects to reach Earth before the visible effects arrive, this being apparent at about ten to the tenth or at least ten to the eighth times "c"; (being somewhat less than instantaneous over a distance of about eight light minutes).

Now these gravitons were of great density (but not infinite) traveling with varying speeds and they formed a multi-nodal agglomerate of space energy matter, which in time became stable in the mean across the universe.

What happened next was that the gravitons collided with atoms and specifically nucleons which are theorized to be the parts of an atom that exist in the dimension of the gravitos which allows the passage of gravitons through them. Again drag was the result (and eventually if a graviton passes through enough nucleons it will slow down to zero velocity and be captured by one).

Unfortunately for the atom trying to return to zero k the energy is either re-emitted as BBR or returned to the proton (via quantum mechanics as a quanta level energy change) and the proton gains a rise in universal temperature and so regains more energy to emit which it will do so instantaneously, by re-emitting the gravitons in forms typified by a photon. If a proton receives a photon via the photos it will instantaneously re-emit it if its energy "charge" is full *. (Note: It is probable that gravitons in the fist instants streamed throughout the universe by BBR from destroyed neutrons and hence gravity in a weaker form was already at work immediately after creation. It is theorized that gravitons are the dark matter (currently theorized) and it is this dark matter which is the cause of gravity. However gravitons do not have any attractive or repulsive force effects on either themselves or other objects and cannot therefore be used en masse to solve the galaxy rotation problem, which I will be addressing in a later chapter. *High energy affects on nucleons by high energy photons will be addressed further on.

This begs the question of what happens with more massive objects. Answer. Stars received so much energy via gravitons that they in turn generated sufficient heat to enable fusion under intense gravity, and they began to "shine" as their clouds of atoms were forced together and fusion commenced. Some black holes are clouds of atoms which became so dense and massive that gravitons were completely captured; with all their energy being expended with in the body of gas, which heated the huge star up to the BST and so enabled atoms at its centre to reunite into praetoms and return to a cosmos like state.

CHAPTER 3

GRAVITY, MASS, AND MOTION

The following mind experiment contains concepts which may be difficult to grasp, by even the most brilliant minds. This is because it includes a necessary supposition which is completely foreign to our "day to day" human experience, (Let alone scientific "knowledge").

This supposition must be retained in your mind at all times during the course of the experiment and (believe me) your mind will most likely continually reject such a concept on a regular basis during the experiment.

To obtain an understanding of the results of the experiment we must ensure we keep reminding ourselves of this suppositional concept: Which is; that the objects in the experiment are deemed to have no intrinsic mass, weight or momentum.

First of all I will relate the objects in the experiment with their other properties:

Imagine some archers taking aim at a nice fat perfectly round cantaloupe of even density. The cantaloupe has no mass but is made of "stuff" which will offer friction to arrows passing through it, but magically without sustaining any damage whatsoever.

Arrows: These are magic arrows also which cause friction when they pass through said cantaloupe and they will in consequence lose some velocity in direct proportion to the friction, which is fully determined by the relative velocity between the arrows and the cantaloupe, (which I will now refer to as an object). The arrows have momentum but are deemed to have no effect on each other and can similarly pass right through each other and in so doing do not lose any velocity or direction.

The mind experiment:

The object is said to exist in a perfect vacuum of space and without gravity. A number of archers are gathered at various points around the object, and for simplicity in this experiment they are (unless otherwise indicated) deemed to exist on a flat (Cartesian) plane.

If one archer shoots an arrow at the object (and we will consider the archers to all be perfect marksmen and on average the velocity of their shots will be even). The arrow strikes the object and because said object has no mass it will remain on the tip of the arrow and instantly accelerate to the velocity of the arrow which will carry it on its tip ad infinitum and without losing any velocity itself. However this is not the case in the real world (which is to be explained in due course) and we will consider that both the arrow and the object are massless but exhibit the stated properties.

Now consider two archers; each on opposite sides of the object and they both release an arrow which strikes the object at the same moment with the same impetus. In this case the object will remain stationary and the arrows will lose some velocity by the friction encountered in passing through it, and they both leave the object with a similar and lower velocity in opposite directions (and in so doing would cause less injury to the opposite archer. Ha ha. (Let just imagine that the archers are magical as well).

Now if we gather a huge number of archers all around the object, all shooting arrow after arrow from inexhaustible quivers, and the average period and velocity of their shots is randomly even. Then although the object might be seen to vibrate, it would in general remain in the same position. However it would be observed to heat up because of the energy gained by the velocity loss (hence energy loss) of the arrows by friction.

Now we have the situation that the velocity, and therefore kinetic energy of the arrows entering the object all around is greater than the velocity, (and therefore kinetic energy) of the arrows exiting the object all around.

Now I am going to place another similar object a short distance away from the first one and the same archers continue shooting arrows at the same rate and velocities (such that both objects appear to be one to the archers). What happens is that each object is now in an arrow "velocity and therefore impetus deficit shadow" of the other one, this being explained in the case of one arrow losing some energy inside the first object and so having less velocity and therefore less impetus upon impacting the side of the second object being "shadowed" by the first. The net result when all archers are involved is that there is a net impetus deficit between the two and both objects will move towards each other with the same acceleration until they collide with impulse.

If I now go back and remove the second object and place all of the archers equally on opposite sides of the object and they begin to shoot again, an amazing thing is soon realized. Even though the object still remains motionless, and is still in equilibrium, and similar to the example wherein it was evenly surrounded by archers, it now has the same perceivable property.

If the arrows were invisible it would seem to have some internal attractive force to other objects in similar circumstances. This is what we currently refer to as gravity (or the ability to attract and be attracted by other objects). As you may by now be confident to continue the experiment; let's continue and we should be able to all agree that it now also has an inferred yet similarly effective property called mass. (In fact mass was inferred the moment the first two arrows struck in the second supposition).

This can be realized by arrows being able to cause an observed gaining of momentum as follows. Imagine that the object is surrounded by an army of archers and we add say, another 100 archers into the mix and we get them to stand on one side and fire one arrow each simultaneously. The result is that the object will accelerate in proportion to both the time and the frictional "drag" that the arrows induce in traveling through it.

The object will continue to accelerate until all of the 100 new arrows have passed through it, and then it will have reached a terminal velocity. Even though we might assume that it should be once again in equilibrium because of the arrows continuing to pass through it from all around, it is not, and will not continue to move at a constant velocity. It will slowly decelerate for the following reason.

Because the object is slightly out of equilibrium! The reason can be explained by recognizing that because it is in motion, it is now experiencing a net velocity difference between arrows striking from the front compared to the velocity of those striking from the rear and therefore it has a miniscule extra force of "velocity induced friction differential force" acting against the direction of travel. It would however take a very, very long time to slow down.

Now the drag caused by the passage of arrows exhibiting friction within the object can actually be called a force, and if the flights of the arrows are continuous and equal then so is the force.

If we were to supply an external force with sufficient impetus to accelerate the object to the average velocity of the arrows, we would notice that the arrows behind would be hardly even striking at any significant velocity and the ones in front would be striking at almost twice the velocity*. The force against the front in such a case is many times more than the force of the now severely weakened force from behind, which by logic we can conclude to occur by an inverse square law of force versus velocity. (Or an inverse cubic law because the object is a sphere?) *The forces of all other arrows round the object when angularly vectored would maintain a net sideways force of zero. This means that there will be no effect observed as any spatial direction change.

Now (back tracking a little) the continuance of velocity observed after the application of the force of 100 extra arrows is called momentum and at "everyday" velocities the force acting against the momentum is insignificant and we confer this property on the object as though we consider it to be an absolute invariable.

Keeping in mind that the object has no inherent or actual mass or weight; we can now do several things to the object to demonstrate some laws of motion. First we will investigate methods of slowing it down, completely stopping it, reversing its travel, and lastly changing its direction of travel.

We can either arrange another group of archers to shoot a simultaneous volley of 100 arrows of equal impetus against the direction of travel of the object and it will decelerate at a similar rate to the original acceleration and come to a complete stop. Or we can get one archer to fire 100 arrows once again against the direction of motion, but this time one after another and the object will be observed to decelerate at a slower rate until we notice it to finally be stopped by the last arrow. Going even further with this, we'll give the object the same velocity again, but now let's get say, a thousand (or any number over 100) archers to fire at once or one after the other. Then we would observe the object to at first decelerate to a stop at a particular determinable faster rate, but then we would notice that after stopping it would begin to accelerate back the other way at a rate determined by the rate of shooting and also by the 900 extra shots. In this part of the experiment, by noticing that an object with momentum requires an opposing force to reduce it's velocity at all, we have discovered inertia.

This all explains how (without intrinsic mass or attractive forces residing in bodies that) such properties as gravity, mass, momentum, and inertia can be effectively attributed to an object or body, and that the arrows must continue to be fired in a constant manner to maintain the constancy of these effects* This is very profound if such a thing could be imagined to be occurring, Good news folks; because what is really amazing about this is that similar and constant occurrence of the effect in the universe, can be explained by the theorizing of infinitesimally small and extremely high velocity invisible "arrows" called gravitons. *Acceleration, deceleration and elastic rebound are functions of the vector laws of the forces so involved.

Side inertia on an object with velocity can simply be described in a similar manner, with equivalent side ways vector forces being required to cause the same vector adjusted results of a vector shift in its motion.

This following part of the mind experiment is where the greatest difficulties of comprehension may arise in the real world, and in the past this may have caused rejection of this theory by majority consensus. This is because; apart from the nagging idea of intrinsic mass (which by it's superglue like simplicity) whereby "no known cause" can be declared to be an acceptable scientific conclusion!* other real world observations to do with physical properties of matter enter the mix. In conducting the theoretical mind experiment it must be quite clear that not only does the object have no mass residing within, neither does it have any physical properties expect for friction. *When has science ever been about settling for unknown causes? Why has this explanation of causation led to such opposition? One word: Einstein!

Having made that quite clear (I hope), We can continue to carry out the experiment and then find explanations of how properties of natural objects can affect the outcomes of such occurrences.

On with the experiment then: In observing collisions between objects, effects can be realized by the fact that when the object strikes say another solid and immovable object there can be two theoretical extremes of reaction. It can have an elastic, (rebound or right through) to a fully inelastic collision. The latter is an impulse (splat) collision.

In this case it is exactly similar to the simultaneous striking of many times the normal number of arrows in one direction.

A perfectly elastic collision means that the object rebounds without loss of energy or vector related reverse momentum which is never observed in the real world. That is similar to the previous example except that at the same time as the extra arrows struck on one side. The arrows stopped striking for the exact same moment, and then they continued as per normal. In the real world other forces such as nuclear "strong binding force" and atomic bonds are the forces which take the place of the extra arrows, which are only explanatory.

So we notice that in normally observed collisions in nature, internal properties of objects supercede the far weaker effects of the arrows, and in actual fact the preceding atomic level forces combine to translate to the force laws of acceleration, deceleration and inertia and the conservation of energy and/or momentum, which are the stated laws involved in this case. These laws remain true REGARDLESS of whether the mass is deemed to be intrinsic or caused!

Anyone who contends that the objects must have intrinsic mass because there is a transposition of such inherent "mass" of the colliding objects to energy as heat, are simply proving their own perspective from their own perspective, without showing causation! And because they don't even know what gravity is, as well as not knowing what strong nuclear force or even for that mater, the force that causes atomic bonds is, they are arguing from both ends of unknown paradigms. How can any reasonable argument then ensue when they attempt to disprove a logical model of causation with arguments filled with assumptions and unsubstantiated "facts"? One such "fact" is mass energy equivalence which I will strongly refute. This refutation is actually an adjustment, and also these atomic forces will be later explained by this theory.

Now leaving this for the moment, let's analyze the forces on a smaller object existing within the "arrow velocity deficit" region of a much larger object. The smaller object has fewer arrows passing through it in sum analysis. However it has an equivalent number of lower velocity arrows passing through it from the direction of the larger object, as it does from arrows which pass through it directly from the archers from the opposite direction with unaffected velocity.

The velocity difference still causes the same continuing force in inverse proportion to the density of the larger object, (similar to the second example), and it will accelerate towards the object at a greater rate than the rate at which the larger object will move in proportion to the size (density) difference of the two objects. By simple explanation: This is because the smaller velocity deficit shadow of the small object has a proportionally smaller affect on the large object than visa versa. If the density difference is massive then the force exerted on the large object by the small will be insignificant, but the accelerative force applied to the small object will be far greater than its size would seem to allow. This is because of the large velocity differential of the arrows passing through it being caused by the large body.

If we suppose there are many small objects existing in the "arrow velocity deficit shadow" of the much more massive object, then the force applied to the smaller objects will be seen to be proportional to their density and therefore their effective mass, and they will all accelerate towards the large object at the same rate. i.e. They will all appear to "fall" at the same time.

This whole experiment shows that gravity and the laws of motion can be attributed to an effect called inferred or effective mass by the affect of "arrows" or other particles which possess the same properties to a limited extent.

The laws that effect planetary motion etc. etc. all remain unaffected. The only difference is that actual mass is replaced by effective mass, caused by the arrows!

REAL WORLD ANALYSIS:

I wouldn't recommend attempting to analyze the following should the previous mind experiment remain incomprehensible.

The idea of "particle caused gravity" is one thing; having it being responsible for the laws of motion is quite another, because some of you who are able to get your head around relativity have difficulty in understanding how this new theory works. I will begin by analyzing some reasons that this may be so, and perhaps be then able to render the following theory to at least become more comprehensible.

I understand that controversy is likely to ensue round this subject and that resistance to such theory will be stiff because of the real world intellectual stubbornness to reject the possibility that all ingrained assumptions are not necessarily reality. Also there is the general resistance to world view attacks which can cause a confusion very similar to "pilot disorientation" which can cause brain freeze and result in wrong and fatal decisions. This can occur when faced with instrumental evidence which is contrary to felt evidence which the pilot falsely believes to be the truth and that somehow the instruments are malfunctioning or he panics and simply can't correlate the two in any sensible manner.

In the case of particle theory the instrument may finally begin to be believed if we can just consider for a minute that felt observances and pre-learned mental content may possibly be false and misleading.

Gravity is an extremely powerful force. It only appears to be weak because of equilibrium. Another problem is that we tend to humanize a force as being weak or strong by felt effects. For example if we are walking at just one meter per second and we run into the bedroom door in the dark, the force that is exerted on our forehead appears to be very strong, because it hurts. That force is in fact extremely weak and even on an Earthly scale without mentioning cosmological, it would not even register on the force level meter.

If the Earth which is traveling through space at about a million miles per hour were to suddenly stop. Then you would notice real inertial force.

Centrifical force also appears to be strong to us, once again because of it's felt effects, in fact the force that causes you to experience say ten g is again very very weak and wouldn't register on the scale either.

Using felt force to determine the value of force in the cosmos is at best misguided. Our opinion of velocity is also clouded by reason of low real word velocities that we consider to be fast. We can perceive inertial force from acceleration and deceleration and to us they appear massive. Only an instantaneous impulse event of world shaking proportions would even show a flicker on the force meter.

We feel inertia because of the resistance offered by gravity force particles offering the tiniest resistance force to any movement away from the state of relative equilibrium we exist in. Any object with a velocity which is below the critical velocity of around a million miles per hour is in relative equilibrium. Note: This is but a theoretical and not accurate velocity supposition which is arrived at by real "world" universal observations.

Care must be taken to not confuse inertia with impact or impulse. A bullet fired from a gun has momentum and its target impulse reaction results in the breaking of atomic bonds, (not necessarily nucleon bonds). The velocity was provided by a chemical process. In the case of a rotating sling the energy and velocity is from human biological chemical relations in winding up the sling to terminal velocity. The projectile is being forced to continue to move out of equilibrium until the release, whereupon it remains on the last tangent it was being forced to, and it travels in a theoretical straight line until impact. This immediate constraint to linearity is because even though the forces acting in atomic matter are elastic, the actual force of gravity is 100 percent inelastic.

Inertia was involved in both the wind up and the impact but both were very different, the first was a very elastic inertial event and the impact was a very inelastic impulse event, if say the projectile struck a brick wall for instance.

When you think of the forces at work in the universe, you must consider that any forces that we can notice, even including a nuclear explosion are still only operating around the point of equilibrium, with perhaps only the nuclear bomb causing a flicker on the cosmological force meter.

We may be traveling though the universe at a million miles per hour give or take. The velocity of gravitons is "squillions" of times greater than that, and we exist on the very low end of the power curve of velocity versus drag, which is why the effect of such seemingly significant velocity is not felt.

I trust that regardless of the fact that strong biding force appears to be the strongest force in the universe, you will soon begin to realize that gravity itself is the strongest force and is indeed the cause of binding force which would be quickly realized if a black hole suddenly appeared in your vicinity, whereby all strong binding force would be instantly rendered useless as matter including nucleons flew apart.

If gravity was suddenly removed from the universe everything would fly apart into a fog of quark gluon plasma, ending in eventual motionless equilibrium. By converse logic we can conclude that it is gravity which actually binds everything together within the bounds of their individual constructs.

So with all this in mind; first let's analyze what happens to an atom being bombarded by gravitons passing through it at hyper-velocity* : Take a lone atom (in deep space away from all affects of galaxies) being so affected and you would likely notice a motion of atomic vibration somewhat similar to "Brownian motion" of particles. The pertinent fact is that multiple graviton transitions through nucleons create drag on the nucleon and the average vector resultant sees the atom in a vibrating equilibrium. *Electrons aside for the moment.

Take a massive object and apply the same theory; we now have a multitude of nucleons being bombarded by the passage of an incomprehensible number of gravitons and the whole object has a net continuous force imparted by the gravitons, which creates drag friction and consequential transference of energy (for heat or motion) to the nucleons by means of gravitons losing velocity.

If the gravitons only traveled unilaterally they would cause the massive object to accelerate in the direction of graviton travel to finally equal the terminal velocity of the fastest graviton. However; thankfully for the existence of the universe, graviton transitions are multilateral and even. This is called GD. However and this is extremely important for this theory. Graviton transitions are proportionally more likely near universal bodies than in deep space. This does away with the need for postulations such as "dark matter" which have been derived as being necessary to show reason why galaxies can keep objective integrity. I.e. the gravity in proximity to universal bodies is greater than in deep space and so we can envisage a GD nodal variance in proportion to general matter population.

All objects with motion being subject to GD will eventually decelerate to zero velocity. This seems odd but it takes so long that it would never be really be noticed unless your name is "Voyager"! Yes, it's a fact; the Voyager spacecraft are slowing down! "Mmm; strange is that!" thought Yoda.

Take an atom or a massive object which receives energy by "drag friction" force of gravitons. The graviton force is exerted equally in all directions towards the centre of an object in equilibrium and average motionlessness. However because of graviton velocity loss through atomic (nucleonic) matter you have a greater force on the event horizon or the outer most regions of the object than at the centre and in the case of massive objects as I have explained you now end up with a force all around pushing towards the centre of the object. The greater the density of nucleons in the object the greater the proportional force !

You should by now have discovered the reason for the gravity or mass of individual objects, but contrary to current beliefs) having greater force being applied at the external regions of the object than at the centre. In fact the centre atom would be in equilibrium with the minimum force and energy gain compared to all the other atoms.

This brings us to the fact that (apart from other reasons), all objects gain energy by graviton transitions. (This is why planets heat up from the inside). The effective mass of an object does not significantly change by reason of velocity until the object accelerates to about half "c" () because up to that velocity the average GD differential remains about the same, because any graviton velocity increase in the forward direction is offset by the decrease at the rear in a non linear manner. What happens as the velocity increases then? We will soon see.

To explain gravity by this theory one must imagine the GD of incoming gravitons in any single direction being greater than the outgoing GD (on a stationary body) because of drag and loss of velocity and energy on the gravitons within the body in question. This creates a GD deficit shadow (GS) around an object consisting of atoms (nucleons). As we previously learned the object now has effective mass and gravity. Any other object with a GS of its own, coming within range of the GS of another object creates the combined effect of having a greater GS between the objects than at the vector opposite directions (or simply the outsides) in comparison. There now appears to be a gravitational attraction which if you think it through is actually a gravitational push! The net result is that the objects will accelerate towards each other until they collide.

Objects such as ourselves already in intimate contact with another object (The Earth) have the maximum sum of the Earth's and our own insignificant GD deficit shadow under our feet and slightly more GD above us (All within the solar—Earth GS). so we remain fixed in contact unless affected by the introduction of another force.

The GS of the Sun (although real and very large in comparison to the GS of the Earth by density proportionality) because of the huge distance inequalities appears insignificant to us in comparison to that of the Earth. (There is a point between the Earth and the Sun where we would be in relative gravity equilibrium).

The GD field has been theorized by others to extend for a couple of parsecs beyond a body (which appears to be justified) because the gravitons do not bounce off each other which would result in the rapid backfill of the GD, rather they only slightly change direction by passing through each other and changing direction by resultant vector force. (I must declare that they exist in classical and non relativistic eigenspace).

In a similar way to the previous mind experiment, the laws of motion can be explained by graviton theory as well.

Take an atom in equilibrium. No take an object which is more stable in equilibrium. Perhaps a 1kg object would be good! The object is in equilibrium with gravity and mass both being effected by gravitons.

Because of GD, it takes a force to move the object out of rest state equilibrium. And it also takes a force to effect any changes to an object's instantaneous GD affected state of motion.

A unilateral force applied to an object of 1kg of mass by the drag caused by theoretical unilateral transitions of gravitons at standard GD is far in excess of one Newton. It is actually the same as the force of gravity at the event horizon of a black hole which causes almost instantaneous acceleration. But because graviton impact vector resultant of an object in space (subject to equal GD all around it) is zero. The gravitational force only appears to be comparatively very weak at equilibrium and low velocities. I.E. The force is actually very strong but appears weak because of equilibrium.

As the velocity of an object increases, I would suspect that the drag or opposing force will increase by a non linear possibly squared or even cubed function.* Note: The degree of force can be imagined by comparing the mass of the Earth with the mass of a black hole. If we place our 1kg mass on Earth it will now be noticed to exist in Earth's GS which has a far smaller GS than a black hole. The kilogram only has force acting on it of one Newton. Although the GS of our object may seem insignificant, it is causative of its own GS (gravity) within Earth's GS. *Perhaps a real physicist can figure that out.

With the 1Kg object in rest state equilibrium in space; it now has mass but no weight. If we continue to push it with a constant force of one Newton the object will accelerate to a velocity of about 1/8 "c" at steeply decreasing rate of acceleration. The initial rate of acceleration is proportional to the difference in the mass of the Earth compared to the 1kg object and we know that to be about 10m/s^2. In fact it would continue to accelerate with probably an exponentially increasing resistance until its downstream GS was similar to GSe.

If we observe the effect of the patently weak graviton vector force resultants we will notice (if we want to hang around long enough) that should we remove the accelerating force (supposing a real world velocity) the object will similarly decelerate according to the cubic or exponential function and at the bottom of the curve (which we notice to be, by all accounts flat) it will take a very very long time to bring the object back to zero velocity by graviton force. So the object now has perceived momentum. (However given enough time the object will return to zero velocity and equilibrium). We humans don't have that sort of time, so momentum it is!

Now by calculation let's find out the velocities of these gravitons, and prove the effects (that they are theorized to have) on objects in motion.

If the GS of the Earth (GSe) is caused by GD loss as described: To what velocity would we have to accelerate our 1kg object to attain the same GS behind it (in space)? We would have to accelerate it at g for a time t until it attains a GS which is equal to GSe and in so doing it will reach velocity v which is what we want to calculate. Now GSe is proportional to the density of the Earth (Me) so "Me" is proportional to t. So the time the 1kg object must be accelerated at g for must be the number of seconds equivalent to Earth "mass". i.e.

$$T : GSe$$
and $$Me : Gse$$
$$So\ Me\ in\ Kg : t\ secs$$

To find the theoretical upper limit of the object's velocity at the upper limit of graviton velocity, you should be able to accelerate the object at g for the number of seconds equivalent to; the Earth's mass divided by the object's "mass" which is 1 so it can be ignored. (This is because of matter density dis-proportionality resulting in the GS of Earth GSe being almost fully responsible for g being about 10m/s).

Now "Me" is around 6e24Kg. So we can accelerate our object with 1N of force for 6e24seconds with an accelerations of g (10m/s^2). Assuming the object moves from rest we can apply the formula V=at or in this case:

V=gt
V=10x6e24
V = 6 e25 m/s

Which means: To bring it to terms of the speed of light we must divide it by 3 e8 which equals 2 e17 times "c" which gives us the upper limit of instantaneous graviton velocity. However, because in reality the GD differential will of necessity provide at least a theoretically (at least double) mirroring force in opposition (according to this model), some scientists have calculated average gravity particle (graviton) velocity to be 1e8 to 1e10 "c". This is because of the assumption that graviton collisions on average will cause an average graviton velocity Vg of about half maximum velocity because of the averaging effect of high and low velocity collisions by inverse square law. which is assumed to be about 1e10* times the speed of light). (Note: However the object would be unable to reach this velocity, because of the actual inverse squared (or cubed) force of GD resistance, being explained). I have calculated it to be 1e6 "c". *This velocity is not necessary for any calculations because we have no way of knowing the graviton density. We do know what GD is however it turns out to be the currently known "gravitational constant".

So the effective mass of a 1kg object at "c" can now be determined because such mass is proportional to GD which is proportional to "c", and since the mass equivalent to the upper velocity limit is 6e24 kg the mass at "c" must be about 1e8 Kg. by inverse square law.

Or we can also derive maximum graviton force on the Earth from f=mg

We know that the gravitational force f is caused by GD. Now acting on our 1kg "mass" we have a force of 10 N from (f=mg). In this theory the GD for this example is deemed to be constant everywhere in space, so the GD acting on the Earth is able to be declared because to GD the 1kg "mass" sitting on the Earth is seen as part of the Earth, so we can determine for this theoretical exercise that on average the same force is acting on every Kg of matter in and on the Earth. So Fe=ma

Fe = 10N x 6e24kg
Fe= 6e25N

Now if we remove GD from one side of the Earth it becomes obvious that it now has the whole force Fe applied in one direction and it (the Earth)

would accelerate to 10ms in 1sec. Note: It should be obvious that this is the acceleration rate at the event horizon of a black hole for any spherical AMO. So the assumption that one would be stretched and torn asunder near an "event horizon" could turn out to be sensationalized fallacy.

Now the effects of GD apply to motion so it comes back to how long do we have to apply the GD force on our 1Kg object (assuming only GD on the side causing the motion). to accelerate it to a velocity at which it comes to equilibrium. This will equal the same time multiplied by force result as applied to the Earth. This is the same as saying: How long do we have to accelerate the object to attain terminal velocity by this method, which we can calculate from, a=f/m.

$$a= 6e25N/1 =6e25m/s^2$$

Then we can determine the velocity attained by the object by V=at.

$$V= 6e25 \times 1sec$$
$$V=6e25m/s$$

So by a different calculation we end up with the same result for absolute terminal velocity as in the previous example with no more force of GD being able to be applied, otherwise the object would be traveling at the same velocity as the fastest graviton. So we again have the upper graviton velocity limit.

If we could apply a force (in space) and accelerate the object against the direction of unilateral gravitons, we would notice an increase in the force opposing the accelerating force as velocity increases because of the increase of GD at the front of it. The force used to accelerate the object would need to be increased in non linear proportion to the velocity because of this reason. This is why at and above average graviton velocity "y" the force required to accelerate the object any further would have to be phenomenal to unavailable. This applies to nucleons and all objects containing nucleons at around standard temperature and pressure "STP". (An upper limit of realistic velocity will be calculated shortly).

First let's analyze a motionless object in space. Imagine that the object is moved out of equilibrium by a particular force pertaining to a "number density" of graviton transitions for one second. This would accelerate the object to a velocity which would take an equal number of graviton "hits" in the reverse direction to return the object to equilibrium even if the reverse graviton velocity density was less and the deceleration rate was less. The vector resultant number of average velocity graviton hits in the forward direction would exactly equal the number of vector sum hits in the reverse direction regardless of the time taken in each case to return the object to equilibrium. A simple way of stating this is. Accelerative GD per second is equal to deccelerative GD per second regardless of the actual rate of acceleration and deceleration. Ok that's not so simple? Then it's a bit like blowing hard on a ping pong ball to accelerate it quickly, and then blowing on it softly to decelerate it more slowly.

GD (Gravity) then appears to be a very weak force, but appearances can be deceptive even though it requires very massive objects to declare its existence. GD is actually of such phenomenal density and velocity that given enough nucleons it causes fusion to occur in stars and it even causes black holes, so any small effect applied to a one kg mass on Earth will of course be a very large effect at stellar levels for proportional results.

Because the backfill of GS is random; then it stands to reason that the instantaneous gravitational effect between two bodies at rest is still inverse square law, but contrary to Einstein the total gravity between bodies is seen to be the SUM of the gravitational forces (varying with inverse square law) and not the product. This is one change in a theory which will be no doubt be hotly debated, but the reasoning is that it is simply because the objects do not have gravitational force acting on each other but the same force acting on both of them which causes a "subtractive" GS between them.

Assuming infinity and zero being the limits of mass at infinite velocity and zero respectively, (Infinite velocity and therefore mass are impossible in the universe because in that case time would be annulled); in that theoretical case, we should be able to calculate the mass of any object (made of nucleons) of known rest mass at any real world velocity.

As we previously determined the mass of the 1kg object in space is totally caused by GD with the maximum mass of a 1kg object caused by GSe at "c" being 1e8kg. If we assume an inverse cubic law acting against the linear acceleration of the mass by constant force, (Possibly Pi related with universal bodies because it has an increasing force caused by an increasing velocity (squared function) creating drag per time per distance acting through the whole spherical volume of the object). we can find the mass at any velocity.

If we analyze say an inverse cubic law curve with velocity we notice that up until about half "c" the mass of the object changes very little, especially towards the lower end of the velocity curve where it appears almost flat. This means that if this is the case, astronauts should be able to handle the extra "g force mass" gained up to a reasonably high velocity. (*This is similar to the cubic energy law relating to the flow of air through a "perfect" wind turbine. The GD acts like air passing through the turbine at air volume per second per second). We will calculate this further on.

So our astronaut would notice a proportional increase beyond the effects of acceleration and his movements would require a proportional increase in effort including all of his bodily functions, so it could be problematical and even fatal for him. Also the energy required to accelerate his space ship is proportional to the increase in mass. Further on I will tinker with an idea of how we may be able in the future to overcome these restrictions.

What then would happen to his clocks in this circumstance? And would there be any real time difference between the time passed on Earth in comparison?

No he wouldn't have to adjust his clocks but not because of relativity but because the GS being experienced by his ship is changing in reverse proportionality between the front and the back and so affecting mass, but only as INERTIAL MASS measured in "g"*. Therefore the required energy rate increase to cause the required increase in mechanical and atomic, radioactive, and electromotive forces in the forward direction etc to enable him to keep time would not result in an overall GS change similar to that which a satellite in Earth orbit would experience. So then I strongly contend by this theory, that his age due to a supposed relativity will not change one iota than it would if he was on Earth. *This would mean that accelerative "g" would have to be added to this velocity related "Vg" in order to determine the maximum velocity and acceleration that an astronaut could endure.

We should be able to calculate how inertial mass changes with velocity up to "c", which we have determined without Einstein to be pretty much, the upper limit.

Now I'll restate this as: An absolute GS of zero, and a GD (gravity) constant of about 6.7 kg/m^2 (This is only used in calculations between two universal bodies). together cause an upper velocity limit of around "c".

In the universe at this moment we have a GS that we need not calculate, because we now know that the GD causes a GSe which results in a g force on our 1kg object of 1e8 kg at "c." (Which is not infinite but it is very high for an object that only weighs a kg). By the way at that g force it still only weighs a kg! Its extra inferred inertial mass is only realized by inferred momentum by the dis-proportionality of graviton transitions.

Now the inertial mass must decrease by some non linear inverse function.* For the sake of explanation I will use an exponential function worked backward. So we know that at "c" the 1kg mass has a mass of 1e8 then at ½ "c" the mass is around 12,200kg, at ¼ "c" the mass is 95kg. At 1/8 "c" it is about 12kg so an astronaut would have trouble even traveling at 37,000km/s. At 18,000 km/s his mass would be multiplied about 2or three times. I feel that this might still be problematical and I suspect that an upper velocity limit for space travel is about 10,000Km/s at which his weight would only be 1.2g. As you can see; above about half "c" the exponential increase requires extreme forces to accelerate an object any further.

The next exponential rise above ½ "c" out of 28 velocity divisions requires about 25 thousand kgs of thrust per kg just to maintain velocity. At these velocities "graviton caused" space drag is a serious problem for any potential realistic space travel. Note: any non linear curve will give similarly daunting results.

*I disagree with the Lorentz factor curve, because I can see a logical "faux par" if it is determined that mass should be infinite at "c" and I even have doubts about the cubic function discussed above and I am personally in favor of an exponential function or a Taylor series form, more like a forth power law, and we may yet be able to travel at velocities approaching half "c" without ill effects.

Also; we might consider the energy required enabling the acceleration of particles to extreme velocities. Does that follow Lorentzian curve function or another perhaps? If the Large hadron collider can accelerate proton streams to almost the speed of light then I very much doubt it.

It is a mathematical fact that no power curve can be calculated forward from unity or less, neither can one be worked backward from infinity! Considering that no supporting data is available, how Lorentz managed to do this and have it accepted as fact remains a mystery. If no data is provided then the best valid assumption is really to envisage the line of equality in which the mass doesn't change for any velocity. The calculation should at worst be according to some diagonal "line of equality" in which mass increases with velocity in a linear manner, but of course universal observation immediately rejects this, so it appears that Lorentz used an asymmetry coefficient which "conveniently" placed the curve out of reach of most observations until now.

For any mathematical results to ensue therefore, assumptive mathematical starting points must be generated. It's quite obvious to me that the math has been "fudged" and even worse; such non empirical trashing of logic has been accepted by the academy as scientifically valid. In fact all Lorentz has done is shown by his theory that the assumption that mass becomes infinite at "c" is not provable. Even $E=mc^2$ disproves the assertion! I.e. how can anything have infinite energy? This would mean that accelerating a single particle to "c" would suck all energy out of the universe and into the particle even creating extreme "universal trauma" at about 0.99 "c".

Is this what we observe with particles that are accelerated to this velocity in the GHC? That, according to the famous "curve" should freeze the universe to zero k and create a universal size black hole. Don't panic! Lorentz, Einstein and even Hawking have now been proven wrong!

If you are a relativist with a mind to explore diversity, you should by now consider that you may have been duped by a logical faux par. In direct contrast with this and even though I assume* a curve as well, I at least begin with a mathematical staring point to enable reverse calculation from the upper limit of mass being calculated to be 1e8kg at "c" from a 1kg at rest. *The correct curve should be able to be presented by reasonable scientific process. At the moment I feel that this would not be important, except that notice should be taken of the velocity at which the Earth is assumed to be moving through space. i.e. the GD field, and that no sideways inertia mass anomalies can be noticed by "Earthlings"! So I postulate a Taylor series curve with an extremely asymmetric line of velocity, very similar to the Lorentzian curve, but one which doesn't end in infinity! Much more is to be forthcoming on this subject in later chapters.

Now would probably be a good time to investigate the cause of the laws of motion by graviton theory:

THE LAWS OF MOTION:

What we have so far concluded is; that for all intents and purposes on our terrestrial orb, acceleration rate per unit force, as well as momentum can be considered to be constants. Now with real world velocities in mind we will begin.

Now a pendulum having much greater return force acting on it (IE The gravity of Earth by angular vectors as explained by normal physics) returns to zero velocity in a much shorter time than a similar object in space. Our object is located in space without an Earth or other gravity to help in any way and in this exercise it will appear to have momentum and so travel on, velocity unabated.

As we previously noted; if we attempt to accelerate the object to extremely high speeds we would notice the problem of increasing GS occurring on the downstream side of the object and GD increase in the forward direction creating a drag force in some non linear proportion to the velocity (As mentioned previously, somewhat like drag on an object traveling though air, but not entirely), until we reach a point that we can no longer find enough energy in the universe to force the object to accelerate any faster, and in that case the occupants of an imaginary space ship would be severely flattened along with their ship by extreme g-force. Einstein theorizes that near the speed of light an object would appear shorter. My theory suggests that an object will actually be very very short, with no appearances necessary! (The maximum velocity attainable by an object with nucleons has been theorized by others to be about half "c". I tend to concur).

Now we have discovered the reason for drag in space and momentum. With momentum we know that it requires an effectively equal force in the reverse direction to equally decelerate an object to zero. Side forces and inertia are caused by the same graviton vector effects. Straight-line inertia is simply because without a force acting against the object in a sidewise vector direction the object is in sidewise GD equilibrium. Inertia experienced by an internal "passenger" is simply the physical transference of motional kinetic energy to the "passenger" inside the object having the force applied to, often ending in an impulse collision with the inside of the "carriage".

With regards to weight and mass differences: There will be a slight difference caused by the fact that a mass is being affected in sideways equilibrium by GD and it is GSe that affects weight in the vertical direction. Everything else regarding mass and weight are as per classical physics. It is only the mechanics that are in doubt at the fundamental and atomic level. It must be realized that it was the measurement of weight which led to mass units in the first place; so there is a direct relationship. Notwithstanding this, current mass values will remain constant whereas weight values are able to be observed to change with gravitational circumstances. E.g. The weight of an object on the moon will be less than on earth, however its mass and therefore its momentum will be extremely similar to its earth value.

No other changes to the laws of motion including planetary and/or gyroscopic precession and conservation of angular momentum, the Coriolis Effect, centrifical force or anything else are changed by this theory as it only attempts to explain known law by a different cause because this new gravity theory ties in well with quantum theory as will be described.

This theory has no necessary problem with Einstein's relativistic theory of gravity, (I'll simply allow it to create its own problems) because this theory doesn't challenge gravity per se. It simply addresses a more probable cause of what we observe and ties it all into quantum physic as well.

CHAPTER 4

THE STARS AS THEY APPEAR NOW

Stars and gravity: Prior to analyzing these, here is a summation of the previous, but with some necessary elaboration re the circumstances at the instant of the birth of the universe.

Before equilibrium of any kind was achieved a primordial upheaval of immense proportions occurred. Granted that the cosmos existed in a three dimensional state, it now became rent or (pulverized) by the introduction of space time (universe). This occurrence was not unlike the very rapid mixing of paint but on a four plus more dimensional scale. The universe may have maintained an outer event horizon with the cosmos but because of nodal effects at different points within the universe, clouds of atoms regained sufficient energy to attempt a return to their cosmic state. These areas of "subduction" of universal matter became known as black holes which grew larger and exist to this day (within the observational time slew, of course).

Atoms having less energy than cosmic praetoms were formed in a vast array of nodal regions with the likelihood of atoms with less energy being formed in nodal clouds more towards the edge of the cosmos or in proximity to black holes. *The denser elements were formed further from the cosmos as the area around them lost less energy to the black holes than they gained by instantaneous graviton "attack".

Because of nodal and interfering energy waves some atoms were formed and masses of like atoms appeared. This caused various masses of different atoms to be formed together, and because of atomic forces and proclivities of atoms to combine for the purpose of "selfishly" giving their energy away to other atoms, they joined together to create molecules of various substances. *
It should be possible for large bodies of dense and "element pure" materials to exist in the universe. i.e. large bodies of diamond or gold! And for that

matter, bodies existing of elements as mundane as iron 60 would be surmised to exist.

This is because gravity was now fully in action at an extreme and incomprehensible level, and the atoms and molecules found themselves being crushed together by the net vector sums of GD forces so involved and the heat generated was immense.

Atoms of like matter combined also by the dimension of forces and fields which was also a new parameter which came into effect. The bonds of these forces were not so vastly greater than the instantaneously strong force of GD (gravity) at that moment. Atoms still bound together either as molecules or elemental objects and bodies of varying size, by bonding via atomic forces (electron bonding, valency) and/or by gravitational impulsion, formed objects with massive elemental or molecular structures.

As far as stars go; clouds of hydrogen especially prevalent in spatial proximity to areas of cosmos/black holes lost their atomic identity and became very neutron dense objects and in so doing became forced together yet separated into individual bodies because of the existence of GD null nodes. Many of these became so massive that they slowed gravitons down to zero within themselves and also became black holes. It is proposed by some astrophysicists that there are black holes at the centre of all galaxies and that many more exist that can only be observed by anomalous behavior of stellar bodies, such as pulsars.

To explain this further: At the time, gravitons were so dense that they caused some massive clouds of matter to become crammed into such dense objects of matter that they became so hot they turned into black holes. The internal atoms returned to the cosmos by achieving the BST (at which the atoms become fully ionized) and in so doing they rearranged themselves to enable the return to Praetoms and zero k which is just an "event horizon" away. What I am saying is that the singularity of a black hole is at (or near) Zero k; as is the cosmos. This is contrary to some other theories. I also disagree with Stephen Hawking, in that the singularities of black holes do not emit BBR, and neither do they evaporate!

Paradoxically the temperature of the atoms at the event horizon of a black hole is at BST. This is also the temperature above which, time ceases to have meaning and gravitons and photons are never again emitted. If the singularity is at zero degrees k then, all other dimensions other than the three physical dimensions would also vanish or in the case of the dimension of time, even though still existing, actually become redundant.

Other less significant clouds of hydrogen, along with other gasses and elements lost too much energy and had insufficient density to regain enough by graviton bombardment to become black holes. They simply became dense and hot and began to shed so many photons and other energy in an attempt to entropy to zero k, that they created an effect called fusion in which the new elemental atoms "plasm" to such a degree that even light fails to propagate via the photos and turns in random directions, which inside a massive "star" means a greater trapping of energy. All the while the star is generating heat by graviton "attack".

At the same time it is slowing the gravitons down so much as to create a massive GS and hence a huge mass/gravity, therefore it continues to gain and retain sufficient energy to maintain the fusion within it. (This occurs at the instant of fusing). What this means is, that even though the forming star reaches an incredible temperature and density, it can't actually be seen until the moment of fusion and it bursts brightly onto the scene.

Unfortunately even as you read this it is a fact that stars are either using up their fuel and will one day run out or be swallowed up on a grand scale by black holes. This may have serious implications for some future generation. Because of observational time slew, it may be somewhat disconcerting to consider that the future may have already caught us up.

Stars also affect their own gravity Note Even though stars slow incoming gravitons down; they also create a significant GD by light collisions within and in proximity to themselves as well as by BBR of gravitons.

Let's consider our Sun: Because light being emitted by the Sun in the photos is almost parallel, the high velocity GD created is not enough to back fill its GS to any significant degree by the time it reaches planetary distances. When solar events occur however, graviton production is increased and is measurable as gravity anomalies here on Earth.

This brings us to observation of the health temperature of the universe. The only way we could take the pulse of the universe up until now was by direct observation. However all photonic observations of the universe are in the past and dare I say it; mostly the very distant past. Therefore we have no known way of observing the distant universe as it exists, period! Well that seems to leave us pretty much in the dark really, doesn't it?

Fortunately by my theory there are other ways to "take that pulse"; i.e. One is by gravity, because most gravitons are still traveling at speeds vastly in excess of the speed of light. (I conclude that if photons of light can make it through from distant galaxies then gravitons must also get through unscathed or at least half the speed of gravitons arriving on Earth must be at least about "c"e3 which gives us a far more recent look at the health of the universe than photonic or emr observation). We can then possibly observe in terms of thousands of years of delay rather than millions or billions of light years. The fact that graviton GD at a universal level appears to be declining in our window of observation is probably fair reason for concern. (See Chapter 5)

After the big bang (shattering), energy escaped from the universe and right up until now and is still continuing to be lost. As you are no doubt aware, the third law of thermodynamics also indicates the eventual demise of everything. This will probably see all that is left behind ending up at about the Bose Einstein condensate level a couple of degrees above zero k. But good news for now however: The distant universe could still be humming away at a slightly pulsating level of equilibrium and that is why we can expect some variations of constants including GD.

This equilibrium is caused by the theory that light has the same proclivity to soak up gravitons as it does to emit them. So light traveling through the universe on its tines, emits and absorbs gravitons such that anyone with a grade three education will see that an equilibrium can be reached if the energy cycle is circular and not subject to other external interferences and able to be self reenergized. However the tentative equilibrium can only exist as long as there is enough light to do the reenergizing. Gravitons are dependent on light in the long run and atoms are still attempting to entropy themselves or vanish into black holes and the cosmos.

When the Stars run out of fuel the first thing we will notice is a reduction in gravity which unfortunately we may already be seeing the beginnings of. So it behooves mankind (if he wants to save his "bleep") to not only look towards universal travel but perhaps even beyond cosmological travel!!

Enough of metareligiophysicocosmouniversiosty: Hey is that the longest word ever coined?

I apologize for the vain digression, but I simply have to check! Antidisestablishmentarianism . . . Hey; doesn't even come close!

OK enough levity: So now we have a humungous universe but how did galaxies end up looking like they did? and how did the spin get imparted to them as well as the planetary systems and planets even?

Pretty simple really: In the first instant of the universe everything was really out of "whack": A bit like paints being initially mixed. Graviton GD was very nodal, as was energy emission through other dimensions. These nodes or interferences (which are a common feature in physics) caused lopsided force vectors which began things moving with some spin in certain directions and you know what momentum is about . . . So as objects with angular momentum approached each other by graviton induced gravity the law of the conservation of angular momentum as well as the other laws of motion and energy/work/time came into play. Note: Some of the most distant galaxies able to be observed in "time slew" have not yet formed shapes that we see in closer galaxies.

As far as galaxies go: The "elasticity principle" of accelerative—deccelerative interactive forces has come into play such that as the objects (being stars) were pushed together, it happened by interaction with elastic interference which as we all know creates patterns. Any number of patterns or nodal results is possible without known eigenvectors/forces etc etc.

(No big deal) just practical physics on a universal scale! The flatness of galaxy accretion discs has already been acceptably explained by physicists so I have no need to repeat their work.

How about binary pulsars? Oh no not them! Oh yeah why not? Binary pulsars are deemed to be too big to be revolving around each other as they do. They have a "Kepler problem" and so they should just do the right thing by everyone and crash into each other already, considering as how they have so much intrinsic "mass". They seem to be either defying the laws of known science (or both stars could possibly be orbiting a black hole). However binary pulsars often consist of a smaller and larger body exhibiting wobbly orbits with precession.

Precession is postulated to be caused by somewhat "closed" energy systems being affected by their attempts to maintain angular momentum and energy conservation with slight overall losses being realized which causes the whole ellipsoid orbit to attempt to maintain the conservation and in so doing it is perhaps not always preceding but perhaps lagging! Precession might also be caused because the orbital or planetary system is itself orbiting a galaxy centre.

Either that previous explanation of pulsar activity might suffice; or perhaps by this theory binary pulsars are creating vast photonic and gravitonic emissions which are colliding head on in such a way that the GS between them (angularly corrected) is severely reduced by vector resultants and the gravity between them is vastly reduced. Two such objects not emitting to such an extent to back fill their own and each others' GS would otherwise have crashed into each other long ago.

This is also the probable reason that large massive low energy emitting objects do not appear in close solar or stellar orbits because they do not emit much energy to so cause a GS backfill in the same manner. The large planets which have been observed in the universe in close proximity to their star may actually be emitting large quantities of infrared and gravitonic BM radiation. So as per these explanations, it would be a prediction of this theory that larger light or other photon emitting stars (i.e. gamma, x-rays and cosmic rays) would exhibit far less gravity than they should because of a higher level of backfill of their own GS. This means that they would be actually denser than they appear.

This creates difficulties with both Einsteinian and Newtonian gravity models when analyzing different types of "heavenly" bodies. It subsequently also creates difficulties for space time warping theory and Einstein's ring effect being caused by the bending of light by gravity. My theory overcomes all of these problems and addresses the otherwise "sometimes strange disobedience" to those laws which has been well documented.

This theory of graviton induced mass and gravity may explain difficulties with the planetary observances in our own solar system.

This all implies that our own Sun is also denser than we accept because it's "light" and BBR, (like all stars) affects its own gravity. This can also be supported by the fact that gravitational anomalies are currently being caused by solar activity. This can be noted by the observation of gravitational information arriving on the Earth (and very significantly for this theory) almost instantaneously, compared with the 8 minutes taken for the photonic data to arrive. This not only lends support to this backfill theory but also strongly supports the ultra high graviton velocities calculated herein.

The gravities of the Earth and the planets are also being affected by the continuous graviton absorption into atoms by photon absorption etc. This not only helps heat up the planets (especially the crustal regions) but it also would have some affect on their gravities in an inverse square manner over distance. The solar photon generated "extra GD" is miniscule when evaluated in proportion to actual GD. However in planets on the inner orbits it could be significant.

The overall effect on Earth will be a slight change in GSe, and the Earth will orbit at a greater distance than it otherwise would. This all means that "large body" gravity effects may be quite subjective, and as we have seen will vary from body to body. This will result in miscalculations of the mass of bodies and lead to anomalous precession and orbital observations, which I suggest will not always accurately track Einstein's predictions.

A little summation and further explanation: The universe and atoms are energy machines on vastly different scales. The universe began not so much as a single void in the cosmos but rather a fragmented void with hundreds of millions of fragmented cosmic pieces existing within the universe called black holes with trillions upon trillions of other sub black hole density fragments.

Because of the sudden outburst of energy and the extreme velocity of the energy burst, much of the matter and energy of the original universe reentered or "bounced" back into the cosmos either through black holes or to the surrounding cosmos proper.

The universe continues to pour energy back to the cosmos but having now reached a tentative equilibrium in comparison with the early universe the rate of energy and matter return has been slowed dramatically. This is due to three effects. 1/ The speed of light has slowed to a relatively constant level because of the mean relationship between light and gravity being in current equilibrium. i.e. The atoms remaining do not receive the cosmic "GD command" to emit light any faster. This will remain so while there are enough opposing light collisions and BBR to maintain the GD status quo, by the reactionary status of the Eos. 2/ The rotation of objects around black holes slows down the rate of attrition into them by the angular extension of the distance of travel required to enter the singularity. 3/ The starlight colliding around a black hole will infill its GD with gravitons and cause it to have less apparent gravity than it otherwise would and hence reducing its ravenous appetite.

The reason that so called red shift exists in distant galaxies is not necessarily because of their retreating motion or any change in light speed with observable time. Rather the protonic energy released was probably more in the red and infrared region than anything else at the time of their formation. (This could only occur under conditions of "other world" like gravity compression to allow fusion at much lower temperatures and the photons emitted were subsequently of lower energy because of reduced quanta separation level states within the atoms). If this is not the reason; then it becomes more likely that those same outer galaxies may already have accelerated themselves back into the cosmos by now.

The reason that the outer galaxies appear (in slew time) to possibly be accelerating away towards the cosmos could be because they my actually be being pushed by gravitons towards the massive GD existing at the event horizon of the cosmos in particular. Many may already have crossed the event horizon but not yet observable by light observation or gravity observation because our observation window is both insignificant in time and of course time slewed as well. However this would have to be consistent with there actually being such an event horizon. This remains unclear.

The reason for swirls and rotations observed to be commonplace in the universe is not because of any coriolis affect, rather; because of the massive nodal and interference patters of energy which occurred during the birth of the universe.

As we just analyzed, objects consisting of atoms and clouds of atoms were torn apart and were forced together by gravity. Overall force differentials acting on whole galaxies as well as star systems caused overall movements then in particular directions. The objects moving in relation to each other were under the control of gravity and acting against centrifical force but they also now had angular momentum. Future forces acting randomly upon them in random nodal and interference patterns caused the galaxies and star systems to rotate and form spirals due to the law of conservation of angular momentum. Equilibriums were found within these systems as the universe slowed and quieted its upheavals during the process of moving towards its hopeful state of equilibrium.

What we are left with is only what we see via skewed observation. i.e. Many spiral and other similarly "angular momentum shaped" objects. Standard theories of physics fit the shaping of galaxies as mostly flattened objects spiraling around a central dense matter cluster or black hole.

Black holes return matter to the cosmos usually at random intervals, as spiraling objects of matter are inexorably attracted to the absolute GD of the singularity, and reach the event horizon BST. When this occurs atoms lose their strong nuclear force and recombine in cosmic configurations with parts of other atoms especially nucleons which already consist of "cosmic stuff". In the confusion and shambles of the attempted swirling reintegration into the cosmos by the reformation of praetoms you would find left over universal matter and gravitons swirling towards the singularity as well.

Just before photons become totally reintegrated they collide with an immense number of other photons in a spatial manner which is determined by the GD shape of the whole galaxy and accretion disc which subsequently causes a disc shaped GD. This in turn causes the matter to move in a flat spiral manner.

The force on photons caused by the incredible density of protons is very powerful. Such density causes light to switch tines and begin to spiral also. This causes photon collisions to occur in a particular spiraling manner such that gravitons so produced stream away from the singularity at summative right angles to the flattish singularity (as well as into the singularity).

These gravitons act on left over matter and they restore the GD balance to the side of the singularity and actually reverse the gravity at the centre because of graviton depletion there at right angles to the accretion disc, a bit like a whirlpool in reverse*. Because the GD has been caused by spiraling light the matter at the centre is quite often seen to spiral or flare as streams of matter particles. The matter (consisting of plasma, cosmic, gamma, alpha and all manner of particles) streams out into space at swirling and or flaring right angles to the galaxial plane of the singularity at very high velocities approaching the speed of light. It is theorized that plasma and extreme light effects causing a massive GS anomaly allow atomic matter to travel at such speeds (The eigenvalues of the graviton vectors would change little but the eigenvectors would probably be calculable if the light eigenvectors were actually known.

Powerful magnetic field affects would also contribute to the shape of the flares.

The eigenspace at the side of the black hole is bilaterally single-sided. I.e. The matrix cannot be determined to exist within the singularity). * If this gravity reversal didn't occur then no matter could possibly escape from the singularity as is often observed let alone at velocities of such magnitude! In objection to this theory please explain how else this can possibly occur with the current model of gravity. Also how can even gravity itself emit from such an infinite gravitation anomaly according to classical physics?

This flaring matter is expected to be at temperatures below the BST but most likely still at many tens of millions of degrees and it emits vast quantities of light from the plasma stream in all directions (which is how we can see it). Universal and cosmic matter may be reunited in these flares to form new universal matter (even consisting of some of the more dense atoms of the periodic table) as the plasma cools down and atoms and molecules reunite.

It is most likely that black holes in the centre of flat galaxies with accretion discs positioned in intimate proximity to the event horizon would actually then probably exhibit a do-nut shaped singularity. This shape is the most likely form considering the lower GS exhibited in the planar direction of the disc, and also the very low GS in the direction of and centered in the flares. This would allow the continuation of the flares which actually would have begun flaring before the black hole even matured.

Light is deemed to be severely restricted from leaving a black hole in multi spatial directions only because of the massive bending (by tine shifting) that occurs because of the massive praetomic force which is thought to be even greater than universal state protonic force*. This is the reason that light as well as ejectile matter is a major constituent of flares. *This has nothing to do with the supposed intrinsic and super gravity thought to be possessed by a black hole, which according to this theory has no affect whatsoever on photons! Otherwise please explain how light can suddenly go from velocity "c" to velocity zero, within an infinitely small distance. Once again please explain how matter can escape the black hole from the observed intimate contact.

With the standard models of black holes, flares are really a conundrum without any possible cause and they should not be expected to exist. However with this theory, ejected matter is allowable and is therefore able to be expelled from the central much lower GS centre of the "do-nut". It is rejected because it consists of particles of matter which (individually) in the threshold instant are rejected by the repulsive force (already theorized) because they have no reunification constituent matter to recombine with to form cosmic matter and are rejected at "c"

Of course they will recombine with the other matter just outside which we have already stated to be escaping the event horizon and will therefore slow down to less than "c" by the energy loss of inertial change by the change of direction so forced by the recombination, and they will exit the vicinity of the black hole via the flares.

This recombination upon colliding allows universal matter to reform upon cooling of the flare in a chaotic and therefore non consistent manner. In so doing lumps of similar elemental and molecular matter will be formed because such occurrences are not constant but very nodal in nature.

It is also a definite assertion of this theory that the density of praetoms or neutronic matter inside a black hole is not considered to be much denser than a neutron star. The only difference is that a black hole appears to be very dramatic because it is in the process of having matter steered into it. And a neutron star is not. The idea of gravity, atomic or other matter densities greater than what the laws of the universe and the cosmos allow is ludicrous.

With non observance of relative size of course any arbitrary point can be of any arbitrary size, so point size then, fails to have any significance, and allowing any (and even universal size) amounts of matter to ever exist or have existed inside of a single point is simply "mind games". It may be fun for physicists to attempt to demonstrate intellectual prowess with the stuff of the mind: But that is not science! Really?

Because pattern and interference nodal effects are common in physics, massive solids of such atoms could well be produced and under the forces involved being swirled like mixing paint into objects of more or less dense matter as the case may be.

Black holes can theoretically be of any size consistent with the density of nucleons being large enough to create a GS of 1. (This becomes the case by there being absolutely no gravitons exiting the object in vector resolution). This enables the singularity to have sufficient matter pushed towards it to enable a GD of zero to occur at the exit side of the event horizon which causes tremendous and sufficient one way graviton traffic to occur through the nucleons of the atoms at the event horizon to cause them to reach the BST. This is because for one reason light is disabled (because of reasons just explained) from diluting the incoming gravitons and a compounding reason is because of the fusion temperature realized in the stellar material being forced into the even horizon.

Anything less than this and the object is not a true black hole and may even become a neutron star and perhaps never be able to reach the requirements for black hole qualification*.

A Neutron star or magnetar consists of a tightly packed mass of constrained neutrons which forces their magnetic dipoles to line up which creates an intense magnetic field. This I turn causes stresses on the body which defy the imagination and magnetars are in the process of beta -ve decay in massive spurts. This immediately results in electron positron (proton) decay which results in a massive stream of biracial gamma photons, a large proportion of which remain joined as magnetic dipoles. These have been observed to compress the magnetic field of the Earth because they have been dimensionally shifted. A direct hit from a gamma beam from a magnetar cold have serious implications for earth dwellers.

Quasars are massive black holes existing on a galaxy eating scale. Black holes can't be seen only their effects are seen. A black hole that has completely swallowed up its own galaxy may never be observed. Don't worry! If one was approaching us we would notice severely increasing GD anomalies and we would have advance warning of impending doom which of course we could do nothing about. * If GD were to decrease dramatically then black holes could lose black hole status and explode back into universal matter in a similar manner to the initial explosive events that occurred on a universal scale at the instant of creation. This may have occurred many times in the past, but it is not likely to occur any more for reasons which should become clear in a later chapter.

Neutron stars are stars that have collapsed to maximum density on the verge of becoming a black hole but have no matter left surrounding them to "suck" in*, so for the moment they remain on the verge of singularity status and could be actually white hot approaching BST but unable to achieve it. The internal atoms have collapsed to form a mass of neutron ions, (if such a description is permitted, this would be because they have no charge in that the protons have actually reabsorbed their electrons to become neutrons). of cosmic like material. Even though they would have no nuclear bonds or atomic elemental or molecular structures they find themselves unable to form praetoms for lack of energy but they still emit some photons as cosmic or x-rays in a similar manner to black holes as they are being hammered by gravitons (but not stellar matter), and they exhibit a massive GS. If they rotate they can emit beams of photons in a lighthouse like effect. They still remain in the standard photos /gravitos equilibrium state like the rest of the normal universe. *Just a gentle reminder that we may have to learn to think "push not pull" Myself included.

Background temperature differences noticed in the universe show the chaotic nature of the shattering and tend to prove that the universe did not arise from a single point of existence, which to my mind is a difficult postulation to come to grasp with and by this theory; unnecessary!

CHAPTER 5

THE UNIVERSAL PRECURSOR OF
EINSTEIN'S ENERGY LAW
DERIVED BY THIS THEORY

The task of calculating formulas for a four dimensional universe is daunting. To expand calculations to more dimensions is a far greater challenge to say the least. So please bear with me as I attempt to explain the almost inexplicable.

In regard to Einstein's "energy-mass—distance per time" law: First we will analyze the perfect universe where energy "out" and energy "in" move instantaneously and there are no losses to the cosmos and the energy divergence number is zero. In this case the universe would be instantly and constantly stable because the infinite number of photons would create an infinite number of gravitons that by instantaneous photon emission-absorption divergence would be evened out to equilibrium. However as we all should understand, energy is lost in a non-enclosed system and the same goes for the universe. As I have stated before the universe is headed for entropy but there appears to still be a tentative equilibrium at the moment (and if we can ignore slight changes in the cosmic pulse), hopefully for the foreseeable future.

If there are losses to the cosmos the only way for this state of stability to be achieved and for divergence to remain zero at the moment (and even should photonic energy decrease by stars running out of fuel), is for there to be a change in the RATE of energy divergence across the universe on either the energy "in" side of the cycle or the "out" side or both. If this couldn't occur the universe would be in a state of permanent decay with the mass/weight/gravity/"c" continuum degenerating at an observable rate before our eyes.

As it turns out photonic energy as well as all other forms of energy are being lost to the cosmos without any significantly noticeable decay effect in real time. This is because of the fact that photonic emission absorption divergence is remaining at zero.

This resulting equilibrium is because there are still sufficient photons colliding in the universe from light being emitted both now and long ago to do the job of keeping the energy cycle going.

Now the energy cycle consists of photons going out and gravitons coming in, less losses. Looking at the inward energy side of the energy equation: "y" (velocity of gravity) is determined by cmf which at stable GD remains at a constant average, and there remains a resultant declarable constant Ed (divergent energy). This is in consideration of the fact that the energy in the out direction must be continuing to remain constant.

Affecting changes to Ed is therefore totally dependent upon the energy out side (photons) which is not governed by a constant motive force. This allows us to calculate the energy out Eo minus energy losses to the cosmos by all pathways Ee. The result is Er (resultant energy out) Er=Eo-Ee

Now the instantaneous quantity of Ed in equilibrium is equivalent with relative terms to the instantaneous mass state energy of the universe*. Therefore if Ed is seen to be constant then Er must be only an equilibrium state constant, which has arrived at equilibrium by changes in the RATE of Er transfer because gravitons are fully causative of Er but photons are only partially and belatedly causative of Ed. (Remember the divergence number zero is not an actual number defining energy. It is simply a number defining equilibrium, so the Ed in the formula is not zero it is actually a real value of universal energy albeit with constant losses being proportional to GD which is currently stable, so Ed still remains a constant (only in equilibrium) with time delayed stability totally reliant on Er). *Keep in mind that divergent energy Ed can only return to an effective instantaneous zero divergence number and not go negative (Ed=Er only at zero divergence).

Note: This means that their was an initial rapid resolution of divergence disparity (by power law), which by necessity requires a much higher initial velocity of "c" and even "y"*, ending with a slow leveling out period which we currently observe here on Earth as steady state "c". *Slight changes to "y" would be deemed to be irrelevant for this calculation because of its already extremely high velocity. This however would have to be factored in for any more complex and accurate formula. Measurements seem to be rather problematic at this stage.

Now Ed has a proportional affect to cause mass and can be stated to be equivalent to effective mass with appropriate terms. So we then have the formula.

Er = Ed*c^2 or being transposed Ed = Er/c^2 being stated: Resultant energy (an obviously phenomenal value) in the universe is equal to divergent energy (mass) times light speed squared. (The squared function is simply because we have a three physical dimensional universe and not a flat one).

The favorable result of this would be; that if any energy state changes in the universe, there should be a change in the rate of energy transfer by instantaneous Er change to maintain energy equilibrium. However the facts are that "energy takes time to move over distance." (We can call that a law!) But in observing the size of the universe it is obvious that it would take a great deal of time. So it is clear that since creation, time has been conveniently employed by the utilization of a decreasing "c" from an initial very high velocity. Because of reasons described in this book (even with a decrease in "c" and because of its steady state velocity) light travel lag enables the equilibrium that we now enjoy, in that we exist and life goes on.

If in the very distant past some action occurred to change the cmf, steal some GD or turn off a heap of galaxies then we could be in real trouble because although we might expect that the rate of energy transfer to the universe would need to change by an appropriate increase in the speed of light, it turns out by the formula that "c" would actually decrease even further and compound the problem in the long run. In other words the universe is dependent on there being sufficient energy moving through it, full stop; and it is the delay of divergent energy being transferred being caused by a now constant "c" which is buying us time. How much time is an unknown because we don't know how long the universe has been in the state of equilibrium and we can't really SEE out there, can we?

The end result of all of this is that because the velocity of light at the time of creation of the universe was almost infinite followed by a decay curve as gravitons began to fill the universe to achieve stable GD we can deduce that the universe can be nowhere near as vast or as old as we assume. (This assumes that we are not yet seeing the light of the most distant galaxies yet, or if we can; then this fact cannot be determined and the question remains open). Apart from this it is reasonable to suggest a much younger universe and that if we are measuring any slow down in the speed of light and any change in the energy output of the Sun or any reduction in the measurement of weight on Earth then we should be concerned that the end game may already have begun.

So what we can deduce is that the real energy formula Er=Ed*c^2 worked well to bring the universal divergence to an equilibrium even though it is only true at instantaneous velocities and at the moment the calculation is modulated by time and calculus must now be used to calculate the rate of energy transfer. However I cannot see this achieving any useful purpose for now.

The universe now has the appearance of stability even with the extremely long time or rate of Er change currently realized. However even the formula E=mc^2 does indicate that if E (Er) decreases because of a reduction in energy fuel in the universe, c^2 will eventually see a proportionate decrease as well which it may eventually do after many billions of light years to be further decayed by a decrease in m (or Ed). (Note: In slew time the billons of light years could possibly mean tomorrow! And we must also consider that at the current time any loss of gravitonic traffic from here on in will result in a change to effective mass that can be seen as irreversible (even if "c" could again reach high velocity). The effects of any significant change in GD may be too horrible to contemplate!)

Therefore "E=mc^2" and its multi dimensional variant "Er=Ed*c^2" are formulae that are only accurate at one instant; being the instant of the big bang, because the divergence component injected by time delay causes the relationships in the formula to lag over time, now being assumed to be billons of light years, so the formulae are only useful in showing that universal energy is actually non sustainable over infinite time.

E=mc^2 is only perfectly true in two states. The first being zero equals zero times c? squared. (c? meaning any value) This occurs in the cosmos and in an infinitely small universe or space which doesn't exist. The second state is for an infinitely large universe which also does not exist.

The only thing that will enable E=mc^2 to have a constant value of "c" in a three dimensional space with a volume greater than zero is if the two other elements E and m are in a stable state of equilibrium. "c" is not the determinant of E and m, rather "c" is dependent on the values of E and m. The whole equation is distorted by time (rate really, because c is actually d/t) when c or "y" (velocity of gravitons) is any number less than infinity in a closed or constantly energized system with said volume.

A summation in simple terms: What all this is saying is. In the beginning light speed decayed quickly as GD increased and this energy increase caused an increased m as per E=mc^2. This brought about a state of equilibrium because of photon-photon collisions resulting in the loss of photon energy (gravitons) along with the equal ability of photons to emit and reabsorb gravitons (being proportional to GD) and hence stabilizing the balance of photons and gravitons to the point where e and m result in the currently observe value of "c". At quantum levels, where e, m, d and t are extremely small, e can be almost determined to be equal to mc^2. To assume "mass" energy equivalence from this is a real stretch!

While sufficient light has been produced in the past to continue to facilitate divergent energy balance, the current state of equilibrium will continue to be observed. If however most of the galaxies that we can now observe in slew time are about to, or have already entered the cosmos/black holes, then we could be moments, decades or centuries away from an unstoppable decay in energy and therefore temperature and mass and the impending non existence of life. This is NOT a doomsday prediction, it simply means that our "thermometer" of the state of things should probably be gravity (specifically weight) rather than relying on what we see with telescopes no matter how powerful they may be. Telescopes can only see what "used to be", a very very long time ago. Gravity is much much faster at delivering the "tweet"!

The problem with seeing weight as a constant is that Avogadro and other scientists never in their wildest dreams believed that "mass" could change in relationship with the number of atoms in a mole.

We have the situation that the mass standard in Paris has been changing and instead of recognizing the "mass change" possibility, scientists are now making a new one. If this keeps occurring then the change in the weight of a standard will not be seen as a change in GD causing a change in effective mass, rather some other reason will be sought. The mass standard in Paris has already lost about 50 micrograms. This loss is predicted by this theory and may be significant if you proportionally calculate the loss of mass of the Sun.

Also by fixing the standard meter to a component of a variable called light, we may be condemning ourselves to never be able to measure any change in "c" because we will be using "c to measure itself. Remember that assumption is the mother of all . . .

Ok duh! So we can remove the mystery surrounding the famous formula if we at first reorganize that "c" is not just some fortunate fluke of nature, but that "c" is just the natural end result of the universe coming into energy balance. (In another universe "c" could be different but the relationship of the terms would be the same). This is because mass is dependent upon energy which in turn is dependent on the DECAY of "c" to the state of constancy which we now observe. i.e. If "c' didn't decay, then the universe would either be far too hostile for life to exist, or it would have run out of fuel long ago!

Now we can perhaps have a less technical but expansive description of the hypothesized behavior of the relationship between GD and light (photon emissions of all kinds) and the effects on the universe and even the implications for the future of the human race.

First of all it must be recognized that photons have the ability to both emit gravitons as well as soak up other gravitons that it encounters. If this were not the case equilibrium of any kind would never have been reached!

Imagine a large GD or graviton density at the beginning of the universe with gravitons traveling at high velocity right across it. Stars are fusing everywhere and light is streaming across the universe in all directions at super high velocity as well. Imagine that the cosmos has been struck and like any elastic object when struck it "rings" with initially high impulse response which quickly attenuates. This could also be likened to a bouncing ball in basic respects.

Now light behaves as we have just noted and GD is very high. In that eventuality light will soak up proportionately more gravitons than it emits and equilibrium of sorts will be reached over time. However this will then cause a reduction of GD which will in turn cause a reduction in stellar fusion which will again cause a reduction in light, so gravity will be therefore pulsing in value and if both velocities remain constantly high you might observe a pulsing of GD measured at first in minutes or so becoming less frequent* to end up pulsating over time periods of years or decades, which if left occurring like this would have torn planets and planetary systems apart. * Amplitude attenuation was caused by losses to the cosmos. Frequency change is caused by the change of the rate of energy divergence which is fully explained in chapter five.

Fortunately for a reason which will be forthcoming in a later chapter the speed of light began to slow down according to a power law and the light travel time across the universe became delayed and the resulting fluctuations in GD became of longer wavelength and amplitude (This decrease in amplitude was because energy was also escaping from the universe). If you object to this light velocity change scenario and ask; where is the proof? How many different colors do you observe in the universe? And also how many stars have been so frequency shifted that they only emit cosmic and x-rays? So I am taking a higher stance to ask; where is the disproof?

This decay occurs until a point has been reached where we notice fluctuating very long wavelength GD pulsations, and this is the supposed "steady" state of equilibrium that we are in today, which we assume (for no good reason apart from recent observation) will continue unabated.

I've got some bad news for you Sunshine! Even if the speed of light has stabilized, there could be another unexpected GD gravity pulse coming our way because we have failed to make sense of the evidence provided (not only by physics) but by geology and astronomy!

In the past these gravity fluctuations have caused planets to shift orbits, changed the rate of stellar fusion and at some state of super gravity the solar system could have been a binary pulsar for a while with Jupiter undergoing fusion for a universal sized "while".

During these periods (and there would have been many of them in the past) Planets would have switched orbits, undergone gyroscopic precession with the consequential internal planetary turmoil resulting in massive polar shifts and magnetic field reversals. No life would have been possible during earlier hot or cold and extremely violent events.

These were the times when planets even collided and Moons were formed. Perhaps woe betides us, if the next pulse is imminent! Let's just hope it isn't.

Past geological history suggests that the last pulse was a low GD "wave" which predicated a colder Sun, which may have only resulted in a minor polar shift but still a magnetic field reversal. Logic then declares that the next half cycle of the pulse will be a high GD wave resulting in a hotter Sun. It might be a good time to invest in an air conditioner manufacturing company.

Of course, if we are still noticing the effects of the last cold "wave" and we are ignoring simple facts like loss of mass in Parisian "mass standards" then we very likely won't see it coming but we can just blame it all on man caused global warming!

To explain all this in simple terms: Imagine that there was no time". In that case the Eos (being torn by two conflicting forces) would instantly reach a state of force resolution. This is because the process would be inelastic. However the introduction of time delays into the situation puts elasticity into the equation, and then the whole process becomes a bit like twanging a taught rubber band.

The Eos has responded to this by reducing the tension on the universal "rubber band" by reducing the speed of light, and this achieves the same result as if you reduced the tension on your rubber band a little bit and you would notice that the frequency of the "twang" would subsequently be reduced in some proportional manner.

CHAPTER 6

LIGHT THE GREAT ACHIEVER

In this theory I suggest a fifth state of matter which is light as a plasma but with no charge or mass. I suspect that such plasma has already been unwittingly created in laser experiments and not recognized as such. Some scientists believe they have created light plasmas by other means. Those are not light plasmas! Light plasmas can be cold or hot but they are invisible. It is only as they decay and photons re-enter atoms and become re-emitted that it may become visible as an effect only if the decay is slower than instantaneous. Light retention anomalies have been noticed in laser experiments, but other explanations have been proffered.

This speculation is not necessary as proof or disproof of this general theory. It can explain however the above effects and such things as plasma balls that have been observed (in one case by a group of incredulous scientists) passing right through an air hostess.

The portentous possibilities of determining the veracity of this speculation is in the theorized proposition that such plasma not only has no "gravity" it may be able to be used to DEFEAT GRAVITY!

Most light plasmas are as small as a photon and are extremely short lived and unable to be observed they are theoretically named gravi-photons. Only plasma created in a vacuum may remain stable. Even though it would be unobservable the gravitational anomaly should be observable and measurable. Light plasma is unable to be contained by a physical enclosure. However the gravi-photon plasma would eventually decay and move back onto tines and be reestablished with velocity by the Eos and instantaneously continue on its way (albeit on a different tine) as if nothing had occurred.

It may be important for the theory that it must be concluded that Maxwell didn't calculate the speed of light. He calculated the speed of "electric and magnetic field propagation which so happens to be "c", which is what convinced Einstein that light must be an "emr" We are going to discover that this is based on two false assumptions and if the coincidence was empirically treated it would have been left to remain as an enigmatic coincidence. Maxwell's formulas then are to do with electricity and magnetism and NOT light, or for that matter emr". This is a supportive conclusion for the following theory of light and emr mechanics.

The idea that *light is an "emr" is only a theory albeit seemingly supported by some observances. However I intend to show that these observances can be explained by the particle theory of light and that light does not exhibit a wave particle duality. *If light is an "emr", How come it is not affected by electromagnetic fields?

The speed of light and the speed of "true emr" are deemed to be set by the requirement of the universal voltage of cmf via protonic action to keep the "temperature" at a constant level by setting the rate of energy transfer through the universe, (or we might all shrivel up and die by either heat or cold!) Light therefore is not necessarily an emr simply because it has a similar velocity.

This theory contends that all photons, including gamma photons exit the nucleus and are not emitted by electrons. Has an electron beam in a vacuum ever been observed to emit photons? Having asked that, I will admit that in a decaying universe, anomalies abound.

Atoms are a disconcerting enigma in normal Euclidian, Newtonian or even Einsteinian physics as the quantum activities often appear to defy the laws and theories therein derived. Within this theory they are seen to be acting normally, as they (atoms) in the main are the cause of light as well as fields and charges, "emr" propagation, gravity and "mass" and as such they are not subject to the laws of effects and objects derived by them.

After gravity I would consider light to be the second greatest enigma in physics. Its behavior under certain conditions can be strange indeed. The greatest problem in understanding light and its behavior is the restrictions also applied by the attempted understanding based on the "three plus one" dimensional universe. Within this framework there can't be a comprehensive understanding of what light actually is, what it consists of and even how it is propagated, (How can an object with no mass or charge achieve a velocity at all?), and what it does, where it goes, what the result of that might be, and even why it has an assumed constant velocity in a vacuum.

Another thought according to my theory is that light is a quantasized packet (string) of gravitons which by graviton definition is a packet of force, which, having no mass can be (at first assumed to be) propagated in a similar manner to the "emr" wave and at the same "wave front" velocity: being "c". There is a force within nucleons able to propagate (gravitons) and photons at the same time over different dimensions. Only massless particles can be propagated without a medium! Wave motion propagation requires a medium in which the wave can compress and expand it by the well known method of wave front motion. (As in sound traveling through air).

Having theorized this I still insist that photons travel in the photos and so called "emr" travels in the propos and so they don't interfere with each other. Photonic energy has a different affect on atoms than "emr" energy. Photonic energy is immediately available for the proton to re-emit it at the same instant it is received. "Emr" however must be transferred by charge/field effects caused by electrons. The propos has a close relationship with the force-field dimension.

Electrons are deemed to have no part in photon propagation*. The role of electrons is in the dimension of charges and fields, and along with atoms and ions it is able to cause electric current and magnetic resultants.

Science continues to realize greater understanding of the behavior of light but physicists must keep changing from wave theory to particle theory in attempts to explain certain behaviors of light. I agree with all the physics currently understood about light except for the behavior of light in media as well as reflection and refraction mechanics !

Other questions about light that don't seem to have answers which this multi-dimension theory seeks to explain are: What causes light to remain arrow straight and to remain the same color across billions of light years of space such that we can teach our children the song "Twinkle twinkle little star"?

When light slows down to enter a medium it is thought to be delayed by so called "atom energy swapping" (which relates to the next question). How does light really pass right through solid and other transparent media? Why does light exhibit double refraction with chiral molecules? Why (if light is delayed in media by atom transfer delays) don't the incoming photons or waves pile up? Why then does some material allow atom energy transfers and pass light and others don't? Why does light have instantaneous acceleration if it is a packet of rest state energy which it must be and therefore should have "mass"? How can it only develop energy and therefore "mass" only when it moves? Why does light exhibit spatial displacement at times? Why does light keep the same apparent wavelength when it slows down through a medium if it is assumed to travel by wave propagation? If light changes velocity compared with its point of origin wouldn't its color change? How can a scientist know what element he is observing if he has no clue to what the original frequency was because the originating source may have been moving when it emitted the spectra at "x" frequency? Why does light seem to have intelligence to try to take the shortest possible energy/velocity conserving path though a media? Why does light appear to have momentum if it has no mass as I contend? Of course many of you have already reached for the "relativity" gun to shoot me down with!

Light has fascinated physicist for centuries. The amount of research conducted with regard to the subject is phenomenal. Many of the questions, some of which I have listed remain somewhat unanswerable with any real consensus. This theory doesn't attempt to give answers due to any perception of the writer to possess greater intellect or reasoning powers than generations of real physicist. On the contrary, Physicists conduct experiments and evaluate mind boggling mental and mathematical problems. I have simply stumbled upon an idea which has its parts lodged firmly in historical thought. I have simply imagined a way of collating these into a theory that seems to answer many of the questions by the addition of other dimensions into the mix.

Precedence for such addition of non physical dimensions has already been set by the accepted addition of time as the fourth dimension of the universe. As with time all of my additional dimensions are of cosmic and or universal stuff. All dimensions that we see and observe in the universe must interrelate and be causative of the ability of the other dimensions to exist. A dimension is not an effect. It must have an affect but not necessarily be affected by other dimensions.

Dimensional interactions and effects caused as per this theory declare the following: "Nothing" is not a dimension because it cannot be causative of anything. (Mind games aside) Three dimensional space has three dimensions which are interchangeable and causative of each other. The cosmos is enabled by those dimensions. Time acts in three dimensional space to enable motion and work to be done at a certain rate. This causes energy to be moved and its characteristics to be redefined.

The laws of thermodynamics relating to this universe are enabled by time. The universe contains an energy cycling system which is caused by other dimensions because there is no other mechanism OF ANY SORT within the universe which can maintain equilibrium if energy exiting from the universe is "one-way traffic".

The fact that we have had thousands of years of observed equilibrium and have not observed a winding down of the temperature and energy state of the universe in any significant manner must be because of the existence of a dimension or dimensions enabling very significant energy return.

We must exist within a gradually converting potential energy system which is non linear and perhaps even subject to a non linear inverse law of decay caused by the interaction of other dimensions returning energy over time and at a reducing rate to the sources of that energy, such that at this moment in time we find ourselves at the fairly flat bottom of the decay curve.

Life on Earth was only possible because of this state of relative energy stability, which should remain fairly constant for the foreseeable future unless a slew time gravitational event is approaching because a slew time observational event has gone unnoticed simply because it is unobservable.

Taking the state of the universe as we assume it to be from our primordial vantage point, this is how this theory sees the workings of the engine of the universe on the macro and micro level.

Being a closed loop energy system means that we must break the loop to start somewhere. I will begin with what we can see with our eyes which is a logical starting point. Light is emitted in vast quantities as photons or strings of photons which are emitted by protons at a velocity determined by the protons. Across the universe there exists an infinite grid of arrow straight time line "tines" for short. Light without being affected by anything else will travel by photon propagation along the tine that it set out on without deviation in any manner right across the universe at "c" velocity, until it may sometime be absorbed by reintegration back to cosmic material or atomic matter or decay by other attritional affects such as colliding with opposing light resulting in the emission of gravitons resulting in a slight attenuation of the photon. This is probably why we don't see a white background at night which we should if the light from the phenomenal number of stars should reach us unattenuated.

There also exits across the universe another dimension consisting of lines called the gravitos. The lines in the gravitos I have called gravitines. The two dimensions occupy the same three dimensional space at the same time because they are not "matter" and it is only matter existing in the dimensions that must conform to the first law of time.

Tines and gravitines account for the whole space of the universe. However they are deemed to have thickness in theory because neither can be infinite in number or density because they are both reasonably full of traveling energy stuff, by reason of which, if they were infinite in number their energy would by necessity be infinite itself, which according to observation and the fact that we are here to observe such; it is not.

To answer the question of why does light (and for that matter "emr) appear to have momentum? Do you remember the little black and silver "spinny-thingy" from your school science class (better known as the Crookes radiometer) that was supposed to show that light had mass and therefore momentum?

In standard light theory of absorption the light is absorbed and transfers momentum to the black vane because it has mass, while the light reflecting off the silver surface is a perfectly elastic collision* which transfers no momentum to that vane and the resultant is that the black vane moves away from the light source. *In the real world there is no such thing as a truly elastic collision and the light should lose some energy to the silver vane.

As we determined in the chapter about gravity, mass and motion, and the subject of the behavior of light, the black vane does indeed absorb the light, but according to my theory the light is a "force stream" of gravitons emitted into the photos, and so the GD on the black incident side of the vane is slightly increased which affects the GS balance. From what we determined before, what do you think will happen? Yes! It moves away from the increased GD because of GD imbalance which changes the effective mass in the black vane and not because the light has mass and momentum similar to explanations of classical physics. The effect remains the same but the difference is not in the photon attaining mass and momentum by velocity, but because a photon contains gravitons with velocity.

On the other hand; as we will see later, the light has zero effect on the silver side because in reflection there is also elastic collision because the photons do not even enter atomic nuclei and there is no net force applied to the silver side regardless of whether light has mass or not and the light loses an infinitesimal amount of energy. So we see that the vane spins in the direction that my theory would have theorized anyway and that is the whole point of mentioning this effect. i.e.

I will continue with light in another chapter but in the following chapter I must first address the subject of Atoms.

CHAPTER 7

ATOMS

Unless otherwise stated the following description of theory is at or about STP.

Atoms are tiny force/energy to gravity conversion machines of amazing complexity and wonder, that have lost their way, but they provide the God code of the "four" proton effected characteristics of matter, Transparency, conductance, magnetism, and emr transparency. (Electricity is caused by the movement of previously described matter). Other properties of atoms will be discussed in a later chapter.

Now atoms are multi dimensional objects and their particles occupy different dimensions. (Remember that dimensions are never invisible to the extent that their subject matter becomes invisible to the perfect microscope, being the "Observer").

This theory states that nucleons exist in the gravitos and consist of (among other things) cosmic particles called gravitons which consist of quantasized boson value sets. Quarks consist of quantasized charge boson sets or color values. Anti-quarks are simply reflections in the branes.

Neutrons have mean constant graviton density per temperature and quantum state, which by itself does not confer them with "mass". Neutrons are stated to exist only in the gravitos and are deemed to be aligned only to gravitines and not tines. Protons however are multi dimensional, and as well as existing in the gravitos they may or may not exist in other dimensions as we have declared. Protons also have very similar graviton density within the gravitos dimension and in that regard are similar to neutrons except for slightly smaller graviton density. I.e. they can be considered to be minus the density of one Electron.

To analyze atoms any further it is necessary to speculate as best as we can what they used to be in the cosmic state. As I mentioned before the

Praetom (for simplicity) consisted of four sub particles with neutronic density held together at zero k by similar forces to nucleons. When the cosmos was shattered by an external event (creation), the praetoms flew apart and in the manner described they shed massive amounts of energy as a mind boggling quantity of particles of every description, and in so doing they cooled until they reached the Bose Einstein state and by that time they had become something else entirely, yet which still retained some characteristics of praetoms. They became atoms.

This is why the helium atom and the element of iron are extremely stable or have very strong nuclear binding forces. In fact any atom with "quads" of four in the nucleus, are stable. This can also explain why the alpha particle is the most common particle.

A helium atom or an alpha particle, are as close as possible to the ideal "quad" to enable preferential return to the cosmos but they are unable to become cold enough or hot enough to do so. (With the exception of a black hole: Not even "deep space" is cold enough!)

At the first instant of creation the hotter spots in the new universe created denser and more complex atoms than the cooler areas which tended to remain in a "quad" such as helium, or nucleon "loners" such as hydrogen of which the universe mostly consists.

The protons and neutrons are basically a unit sub nucleon of praetoms quads that have lost a large amount of energy. In losing cosmic energy they still retained the strong and (in certain externally induced energy states) the weak nuclear force. They developed a charge and via the Eos dimension which helped enable them to maintain a permanent connection with one of the first particles the proton emitted. i.e. an electron.

At the very first instants of creation however, they were at extremely high temperatures and unable to keep any electrons, but what they found they still retained was the weak "cosmic force". This enabled them to form tenuous bonds with neutrons as deuterons and as the temperature cooled and the Eos became effective the bond was replaced by the strong nuclear force. The new "ion" was now able to attract as many (more or less) electrons as there were protons in the new nucleus. And the praetoms were finally converted to atoms.

(The idea that large atoms have weaker nuclear binding force than smaller atoms being caused by the fact that the force is generated in the centre of the atom and decreases with radial distance is considered to be absurd. The true reason is that each "sub quad" has a finite force which is diluted by the vector spread of the forces and the vector sum in the centre of such a force field is greater than the outer regions, (except for a shell of stronger vector forces around the outside of the nucleus that helps to bind it all together). Once the nucleus was formed and cooled, no new nucleons were capable of being readily received into the now fully formed nucleus.

Overall however we can conclude that the binding forces of smaller nuclei are stronger than the larger, for this reason. These forces being slightly different compared to the probable results from $E=mc^2$ are good proof that even at quantum levels this formula is only a guide, but it is the best formula available and works well for the science we have today. Multi dimensionalism may account for slight differences in nuclear bond anomalies, and may require further research.

Protons have lower "mass" than neutrons because they have reached equilibrium less the graviton density of an electron. "Mass" at the quantum level must be determined by $E=mc^2$.

When neutrons exceed the graviton density of half an atom by more than a couple of percent, they reach a "quanta tipping point" and instantly transfer the extra gravitons to protons or if the proton is in a non receptive state by having a "similar" energy state to surrounding atoms, the neutron will reemit gravitons as individual bosons back into the gravitos or Eos as BBR. (Note: The whole inter-dimensional interactivity is probably enabled by the other known quantum level particles. These particles may transect the dimensional borders in non Euclidian space and exist (either in whole or in part) on branes and inter-dimensionally. To theorize this inter-relationship is seen to be as daunting at best, and perhaps may only be resolved with further experimentation.

Many of these other particles exist in protons and neutrons to create forces to ensure their best connection to each other. I am referring to the strong and weak nuclear force. Note: This theory does not change any known particle physics but the graviton along with gluons may actually be one of the hypothesized particles which largely fill the atomic density and other unobservable particles may be involved in inter-dimensional relationships within the atom.

Neutrons have "got it made". They consist of the perfect cosmic matter state to theoretically be able to instantaneously convert themselves to gravitons and release 1.2 MeV of energy whenever the chance may arise. (It only takes four whole neutrons to decay to release enough gravitons to make a praetom). A couple of things prevent this instant decay from occurring in free space and even though they will decay, the time taken is increased to minutes because of the continued energy supplied to them by graviton transitions. This delay plus low temperature constraints prevents the instant recombining of neutrons in space. Such a combining of four neutrons would create a miniature black hole of one praetom energy being by deduction 4.8MeV. The other thing that prevents them from decaying at all is the fact that protons greedily hang onto their neutrons by the strong nuclear force. Although this appears to infer that protons have intelligence, it is simply the proton's obeying cosmic law.

One way in which protons rid themselves of energy is to transfer gravitons by convection (via mesons?) to other protons of atoms in close proximity which are at a lower energy state. Nucleons exist in the force field dimension which enables them to keep in direct contact with other nucleons with which they are able to translocate and promote bilateral energy transferability as if electron orbitals don't even exist. Thus they form intimate "connections" of convenience. This connection is NOT a force and in no way similar to the force that holds nucleons together or atomic bonds by electron interactions. (Note: Nucleons are not being selfish and even though seeming to posses intelligence and decision making capabilities, they actually do not. This manner of "personalizing" dumb particles is simply a tool not dissimilar to the description of straight or wavy lines which don't actually exist. Atoms are simply obeying the laws of dimensions in which they have no choice but to do. What is . . . is!)

The proton dimensional state is determined by the aperiodic space filling of the nucleons which must be different for every nucleus. So the reason for the various properties of atoms comes down to something as simple as the geometric arrangements of nucleons !

Electrons do not normally collide because they repel each other dependant on four variables (usually thought to be three, which may be theoretically correct with a theoretical lone atom in space). which also determine the electrons energy state. Fermi level: charge: magnetism, and "spin' (vibration)). If they did and because of external forces they would simply pass through each other like gravitons.

It is now well understood that electron clouds probably form nodal shapes by "quanta" confinement as well as electron "like" charge repulsion and other complicated electron characteristics as stated above. Uncertainty principle suggests that if the electrons are buzzing about very fast we haven't got much clue about where they actually might be. I do have one suggestion; that they move in a dance inferred by quantum harmonic oscillation theory.

It is theorized then, that the inner electron shells show extreme nodal patterns and the outer shell less so by inverse square law diminishment of point source effects by the diminishing electromagnetic field. However it is also theorized that the outer shell of atoms which if filled, will exhibit a spherical shape due to force equilibrium, as observed by tunneling electron microscopes. (Note: "Tunneling" or "atomic force" microscopes are not imaging by light).

Electrons "snap" in quanta steps to other energy level "orbitals" according to known physics, but by a theorized "three to two" wave phase nodal function, and (If I play the Devil's advocate) I would state that electrons are most probably only "indirectly" utilized as the emitter of light which is otherwise fully dictated by the proton, which emits light as streams of quanta. Electrons do display quanta effects when light emission occurs but this should not be confused with it being any evidence of causation.

When a proton emits a photon it recoils by equal and opposite impulse reaction at 180 degrees out of phase. This dampens its vibration by one or more quantum numbers. This then causes a decrease in its positive charge and an electron/s jump orbital/s towards the nucleus, because of the consequential quantum reduced Coulomb repulsive force. This admits that photons when emitted are also emitted as quantum number packets (streams). Multi level quantum jumps lead to the emission of higher level photons such as x-rays, and the "chosen" electrons will change energy level in response.

Atomic (quark) quanta levels are responsible for band spectra and sub quark photonic quantum units are responsible for line spectra. This gives a photon its vibration frequency signature. The individual photon particle packets are quantasized with photon quanta number per Plank's constant (which is the actual photon count per stream) which results in line spectra which is further determined by elemental properties of the emitting atom.

Under normal conditions the atom will "recharge" and at each quantum point, an electron will jump in result. In both cases the Pauli and Fermi conditions must be right, or the atom will remain slightly ionized until they are, and an electron is then able to move.

Electrons therefore are not deemed to have any other function than electro-chemical and electro-magnetic ones. This is indeed not to understate their importance because without them universal matter would not be possible.

I conclude therefore that electrons are not the receivers of photons. The electron by my theory cannot physically touch the photon as it is not in the same dimension. Rather the proton emits the photon (dimensionally through) the electron orbitals and it may be tentatively understood that these orbitals then exert electromagnetic expulsion energy on the photon. How it is possible to affect a string of photons with electromagnetic energy is unclear, so I must suggest that in the end it is unlikely that electrons have any role in photon emission at all and electro magnetic waves don't actually exist!

The effect of electrical current is a force-field effect that is caused by the motion of atoms, ions or electrons and is only observable in a conductor. By the same token "emr" can only be emitted from a conductor. The idea that an electric field and a magnetic field combining in space will form an "emr" wave is probably absurd!

Under the affects of externally applied high temperatures and/or by externally induced electric charges, the protonic charge and atomic vibration (spin) is seen to increase severely in accordance with extremes of nucleonic vibration caused by (equal-opposite) reactions within it as it attempts to shed energy. When this occurs; the electron nodes are forced away from the nucleus in proportion to the larger,* and more complex "frequency pulsations" of the proton. At first; electrons react by jumping orbitals in a specifically ordered filling manner which is well established in science. (Within the constraints imposed by the four variables that constitute an electron energy state, this is most likely due to frequency and phase changes between the electric and magnetic fields which as the nodes reach peaks will perpetuate nodal peaks in different spatial positions that don't necessarily result in an even and linear orbital jump sequence). *I would normally suggest lower frequencies because of the inverse square law of power versus frequency. This is an example of quantum physics defying the laws of classical physics. But remember the law that states that the objects of the cause of an effect are not subject to the laws of the effect so caused.

As the energy state increases, uncertainty principle comes into effect even on the nucleus because we cannot know the positional movements of the nucleon quarks and dipoles which causes the electrons to move way from the nucleus in any particular manner until at extremely high energy states they exit the atom altogether. This leaves the atom as an ion. If left in an enclosed space the electrons will drift around or act in accordance to external electrical or magnetic influences and as the medium cools they begin to recombine with ions. (This is called "electron ion" plasma).

Atomic bonds are severely affected during this process and the object can melt, or completely vaporize. Note: The protons or electron do not lose their individual charge signs during this process, but the nucleons attempt to reach eV parity by shedding energy. (The dimensions of the Eos and force-field, both possibly being causative are unaffected).

Electrons only move and maintain orbitals and spatial positions completely governed by the forces I have described and nothing else. No particle without a charge and not existing in a common dimension can collide with or affect them in any way.

With regard to the magnos: An atom is not the smallest dipole. An electron is the smallest dipole which exists. It will usually be found somewhere near the crossing point of electric and magnetic field "contact" points being the vibrating electric and magnetic fields of a nucleus, (suspected to be at a 2-3 frequency ratio* and as such could be called the smallest dipole. The electron probably exists of a variable quantity of gravitons and a constant number of north and south magnetons. * A variable 2-3 ratio is supportive of the nodal affects speculated to cause the orbitals which are theorized as occurring in periodicity as a function of Hilbert space sets occupying the space defined within the Fermi layer of individual atoms. This supposition may well be challenged but it has an ally in quantum harmonic oscillation theory.

The various orbitals are imagined to be interference patterns of these wave forces subject to "the four" affects. The electron also has a radial negative electric charge field at right angles to its magnetic dipole. The electron is determined within constraints already suggested to keep its charge oriented toward the (varying) charge center of the nucleus, and also to keep its dipole aligned and oriented with the (varying) nucleonic magnetic force lines. It achieves this on a more or less average basis only. Classical electromagnetic theory does not actually support the quantum theory that a spinning point charge will create a magnetic field. Magnetic fields are only produced by the spatial motion of electrons within a conductor. Transference of classical theories to quantum mechanics "ad hoc" is not a valid scientific method and is one of the basic difficulties with unification attempts of the two branches of the science.

This theory allows that matter and anti matter may be able to be formed at hadron levels. Consider an electron and a positron: Being declared by this theory to be objects without mass they must move without kinetic energy or momentum and then recombine without emission of energy and there can be no resulting mass and the direction of movement of the gamma photon so formed is only determined by field/s charge. However I will allow the formation of multiple gamma photons must be possible from electrons in higher energy states but it has nothing to do with momentum.

The relativity holster should be empty by now as should also be the "cross draw" holsters of energy-mass equivalence, which nuclear physicists sometimes attempt to use with some sort of reverse logic whenever it suits them. That is not a fair sledge really, because science is always a journey and sometimes the inexplicable and downright contrary have to be ignored to enable advancement in any area of better understanding. The general idea is OK but we must be wary of dogmatism.

I assert that the electron has been wrongly determined to have mass because of the misunderstanding of the forces involved in proton-electron observance as stated above.

I suspect that electrons are positioned around the nucleus by the following process. Protons have a pulsation of various and variable electric fields from zero to a positive magnitude determined by atomic parameters. The frequency of pulsation is possibly too high to be measured although it can be observed in RF "bucket filling" in synchrotrons. The magnetic field also expands and collapses at other frequencies determined by energy state parameters. These two elastic forces create nodal patterns, in that, rapid changes of frequency of one or the other causes electron orbital nodes to snap to and fro. Such rapid nodal "snaps" can be instrumentally observed when experimenting with sound wave interference. These quantasized nodal shifts can perhaps explain nucleon quanta level shifts in quantum physics.

The electrons are always positioned in a Coulombic avoidance relationship with each other and will always be found near the region of the junction of magnetic lines of force and electric field nodes.

A necessary digression: This will give them a particular spectrum line which of course can be affected by the Stark and Zeeman, and Paschenback effects because the electron is both an electrically charged particle (in the double negative and single positive manner as will soon be described) as well as having a perpendicularly oriented magnetic dipole, otherwise these effects could only occur if "truly" unreal physics is involved.

The spectra effects are suspected to be caused by retro-charge and retro-magnetic effects transmitted to the proton by the electron proton interrelation, which in turn would rewrite the signature vibration of any photon that the proton emits while in the particular state. Multiple lines are caused by multiple atoms being affected in slightly different ways according to their real world parameters within the field.

This high frequency force interaction causes the electrons (which consist of a magnetic dipole and a -ve charge as a dualistic particle)* to move at very high velocities within the constraints of the fields and interactions with other electrons so described. Uncertainty principle declares that an electron can be at any intersection point at any time but (on average), equidistantly between other electrons. *To behave as it does an electron can be conceived as consisting of one north and one south magnetic particle at right angles to a set (or multiples of a set) of three charge particles in a -+ arrangement giving it a net negative charge and also a charge "pole". Note Charge particles (with color changeability) do not follow coulombs law within the fundamental particle they exist in; otherwise a proton would fly apart by quark repulsion! This is a novel theory but how else can you explain the ability of an electron to change its energy level without a comparative color change capability?

According to this theory, electrons must consist of a lot of particles if an electron positron collision can cause the formation of two gamma photons.

All initial particles are deemed to be gravitons and it is these gravitons existing in other "dimensions" which gives those charges and poles. The electrons (and just as importantly) quarks* exist in Hilbert space sets which can explain quantum numbers, which in turn can explain the major yet extremely stable difference in properties of very similar atoms such as gold and mercury. This can also explain why solids of similar elemental purity don't simply assimilate when brought together. The sets are thought to be initiated by quarks and can also explain the reason that electrons remain in orbitals well away from the nucleus. *Quarks could be thought to occupy a "two dimensional" brane separating several dimensions. This may sound very enigmatic because in Euclidean space two dimensions don't exist, only in pre-Hilbert space. So this is not problematical for branes which posses no size or shape! i.e. they don't occupy space and therefore the quark's gravitons can exist in the same "space time", yet in quantum states generated by eigenvector related Hilbert sets which surround the nucleus and protrude into the orbitals to occupy the defined space of the whole atom including orbitals (being configured by the sets). The quarks may also consist of dimensionally shifted magnetons as charge poles. *For the sake of explanation we will incrementalize the quark to have some physical thickness.

Continuing the narrative:—This positioning of electrons is thought to be caused by electrons sliding along magnetic force lines. This occurs by their need to keep their charge and magnetic pole alignments as close as possible to where the two fields cross each other at or near right angles. This gives electrons a transverse motion. An axial motion is also likely within nodes to keep distance from other electrons. This doesn't cause a quantum number change because the Hilbert sets give them axial "leeway" without any quantum jumping being forced. The velocity of motion is so great ("c?") that we are only able to "see" a cloud. A single cloud or node will be seen to have denser and lighter regions because the electrons (more often than not) occupy the expected regions of the nodes.

The nucleus now has a significant electromagnetic "interference pattern" bond on its electrons and of course the bond is equal from electron to nucleus. When electrons from another atom overlap in orbital nodes from other atoms in electron nodal sharing arrangements by differing "set" positions*, they also share the atomic electromagnetic forces which results in what is called atomic bonds between atoms. This interaction causes the nucleons to have to reset their energy levels and they emit energy *This forces the theorizing that there must be a set covering each of the quantum values (s,p,d,f etc).

It is easy to understand the reason that the lower energy levels are towards the nucleus. Electrons are able to jump levels as determined by the nucleonic parameters just explained, which in turn is mostly determined by the energy state of the proton*. It is "Pauli exclusion principle" for electrons in deciding which one in any particular orbital is destined for a jump. The determination of which levels electrons jump to is a well known process and it needs no further discussion except that in reiteration it appears to me that the mathematical periodicity causing electrons to jump to a higher energy orbital than a higher quantum number position in the same orbital is somewhat similar to periodicity caused by interference patterns. (Just a thought so don't quote me) *This suggests that electrons don't realize a change in their own energy state, they only react to quark/dipole harmonic oscillation differences in the force field Hilbert sets, and Pauli exclusion principle is more determined by positional set information being propagated out from the nucleus than by any energy or color state that particular electrons may possess.

A take on alpha particles: I surmise that alpha particles are a "quad" which is in such a state that its multi dimensionalism has changed and the electric and magnetic fields oscillate at the same frequency but ninety degrees out of phase which results in zero attraction nodes for electrons to exist in even though it still exhibits a net positive charge of some description. If this were not the case alpha particles would take electrons with them and become helium atoms! Also; please explain how alpha particles can escape from inside the uranium atoms and through the lead it has decayed into by "tunneling"? I know it takes billions of years but "time" has never allowed the impossible to occur!

Quantum theory works with many repeatable experimental results, but this only suggests that if the wrong assumptions remain then the interpretation of results will remain the same. I don't dispute the fact that fermions and bosons exist. But whether they have spin, inherent mass and stationary momentum I have my doubts, but only because of this new theory. In any case the standard theoretical approach works for now. I simply declare that effects realized by quantum experiments may be better described in a multidimensional way and by the effect of gravitons passing through particles (as described so far in part) and inferring mass.

Anti-particle behavior may perhaps also be better understood with multidimensions allowing space time coexistence of multiple particles including matter and anti matter occupying the same space at the same time.

Even in quantum physics with up quarks, down quarks and "strange and charming and beautiful" ones etc, the same three enigmas still remain. i.e. What is the cause of mass/gravity, charge and magnetism?

I have taken a stab at mass and gravity. The reasons for charge and magnetism, at the level of the electron and nucleus remain elusive and I can only suspect that they are in fact cosmic and/or virtual parameters as I explain herein, yet we perhaps don't need to explain them at all. Current electrostatic/magnetic theory is substantive and convincing enough. I will be having more to say about "emr" propagation further on.

The so called electron spin moment doesn't agree with classical physics by a factor of two to one, which is very significant. So much so that I theorize that the electron dipole is simply caused by it being subject to the infinite divisive affect of all dipoles that in reducing a dipole to smaller pieces the smaller pieces will still be dipoles. The fact that an electron is a piece of a proton which is a dipole suggests of course that the electron will also be a dipole if it consists of the same matter as a proton. So an electron is not required to have spin or planetary motion. It does not and indeed cannot have angular momentum conservation because it has no mass, which is because it doesn't exist in the gravitos and therefore is not affected by graviton transitions. Magnetons and the role they play will be discussed in the following chapter.

I would consider that nucleons are full of a particular quantity of gravitons which determines their energy state. (Remember that all energy propagation in any manner is the role of graviton movement). However it is the "neutron graviton packet" which exists permanently and totally in the gravitos. Protons are able to have their packet existing across branes to other dimensions, and it is this that gives protons their special features, especially with regard to light and electron behavior, as I have previously described.

To speculate a bit further: I suggest that the quarks could be the "receivers of graviton drag and that it is the down quarks in particular that exist in the gravitos which are responsible for this reception of gravity; so to speak. Mass and gravity are then equally apportioned by the strong and weak nuclear forces. (This requires mass energy equivalence to be an incorrect theory). No other fermions or bosons ever exist in the gravitos. It could be theorized that gluons may perch on branes between the dimensions and facilitate interdimensional interaction.

If a graviton is ever to be observed in a particle accelerator, I would be looking for something that always travels in a straight line but with random velocity differences between individual gravitons and which shows zero deviation in motion regardless of the forces which may be applied unless they collide with each other or a down quark. A photon will be seen to absorb or (If your observational media is quick enough) emit a graviton.

If this is the case we would notice that atoms then have an overall mass by nucleon "mass effect" sharing, and that protons have only half the graviton acceptance of neutrons. This could explain the low mass of the H1 hydrogen isotope and the lack of nodes in its electron shell considering its low (no) nuclear binding force.

The protons at matter event horizons and because of their quark arrangement are able to emit quanta packets of gravitons to tines at "c" velocity. These packets of gravitons are called photons (light both visible and invisible). The photons are also given a component of universal potential energy seen as a rate of ability to do work which is only realized as work being done by the emission of gravitons from their package within the region of travel. This energy is bound up in a way similar in some respects to a pendulum. It is given a vibration (from internal particles "spin moments") at a specific rate or frequency. The packets are held together by containment "shapes" in the Eos dimension which is trying to take them back to the cosmos.

Protons recognize photons by interlocution via the Eos, otherwise the gravitons which make up a photon would pass right through the proton. This is a prime contention of this theory.

When a photon strikes a proton of an atom in its own dimension of the photos it is re-gathered by the atom without any "significant" momentum or other motional force being realized. Its greatest effect by far is that it causes a change in the energy state of the atom and its electro magnetic field is consequently increased and modulated into the electron orbitals at the frequency of vibration of the photon at "c", and at the same moment the proton "convects" the excess energy to other intimately (even momentarily) connected atoms via the force-field dimension, with regard to "Pauli". This causes all atoms so involved to increase the amplitude of vibration and also that that of the orbitals in proportion to the energy received and retained.

This being the case; then light acting on the retina of the eye and spectrometers etc. is actually creating electrical signals to the brain and electronic equipment at the atomic rather than at the cellular or even molecular level. This concludes that variations causing the observability of different colors are by protonic electrical stimulation and not by "light waves" tickling some rods or cones in some unknown fashion.

After this necessary digression I will now re-address the circular energy equilibrium map of the universe. Photons traveling in space collide with other photons and even though they have no mass because they are concluded to be one of the causes of effective mass, they are still subject to the first law of time which states that no two objects can occupy the same space at the same time without the release of energy, so something must give.

It is observationally evident that it is not light's velocity or frequency of vibration that changes, so something else must occur. Light is also determined to not have kinetic energy but it has two other types of energy, gravitonic and vibrational*, the latter being caused by the emitting proton. Gravitons it can lose! Amplitude it can lose, but it can't lose its frequency which is its identity or signature. * Vibration of atoms is not energy. It is a resultant force caused by the internal elastic force interactions caused by the internal movement of gravitons within and through nucleons.

A photon has infinitely continuous vibration at constant frequency because the internal force moments of interacting gravitons within it never change because the gravitons ARE energy and so conserve their own motional capacity. Even when gravitons are emitted and reabsorbed by the photon, the frequency of vibration is not changed. If it does then "c" must change as a consequence. (This requires further investigation).

This is what I theorize: Photons interlocute and at the instant of collision an amazing thing occurs. Photons move almost instantaneously out of each others way. *Not spatially; rather they move into each other yet in complete avoidance by simultaneously shifting into the gravitos, whereby they can seemingly pass through each other and emit energy in the form of gravitons and not lose velocity or frequency of vibration by any consequence. However the vibration will lose amplitude (and perhaps the photon will attain inexplicable rotational changes) and the light will attenuate. (This is all possible in a multidimensional universe and more importantly it explains observed reality, so though difficult to comprehend it is no mind game). *(Note I will admit that if it wasn't for the behavior of light plasma by this theory I would allow that photons could pass through each other in the photos).

The photons are now colliding in the gravitos but (under normal conditions without other affects) they maintain direct attachment to their original tine (via the brane) which they return to and continue along albeit with slight attenuation. The energy that photons release in the gravitos is that they emit a graviton or more at vector angles prescribed by the direction of the collision. (The more "head on" the collision the more gravitons emitted). Photons may also soak up gravitons they encounter while in the gravitos.

This is of prime importance for this theory and determines the contribution that photons make to the equilibrium of the universe by gradually canceling nodal and interference areas of graviton densities (GD) and also engendering photon graviton equilibrium at the point of transition for each photon/photon "collision". The end result is that in deep space we would observe a general and somewhat constant gravitational effect and temperature.

Like photons, gravitons also have no mass for the same reason of being causative of mass and are constrained under the laws of the dimension in which they exist. They are under the law of the gravitos which is under the tension of cosmic motive force. As energy particles they are without mass and therefore are not subject to the laws of motion and they move without acceleration. They are however subject to the laws of time (and so travel at finite velocity) which affect all matter as do vector laws.

The graviton speeds away at incredible speed. Some have theorized (similar to my previous calculations) emission velocities of ten to the tenth or at least the eighth times C). However gravitons speeding along in the gravitos collide with photons, nucleons and each other and being subject to the laws of time and energy conservation they slow down a little when thy pass through each other and because they pass through each other they bend the direction of travel of each other by vector force resultants of the "drag".

Now to complete the energy cycle back to atoms: Gravitons traveling in the gravitos pass through nucleons as I have already discussed and cause drag which is a force on atoms which is either reemitted or converted back to photons or other energy by the proton and the process is ever ongoing.

The mean velocity of gravitons in the universe is also averaged by the sheer number of gravitons and collisions. The mean velocity of gravitons and hence a measurement of gravity itself may be someday be proven by the effect on matter at the event horizons of black holes (man made?) as well as the calculation of the maximum velocity attainable by an object with nucleonic density in space because the effect of gravitons on nucleons is "drag". This theory contends that space is far from empty but is full of graviton matter, (dark matter?) and that all particles are multiple quantities of gravitons or their sub-particles and that greater particles, (being "packages" of gravitons) under the right conditions can be transformed into sub-particles either directly or by matter anti-matter re-assimilation. This suggests that electrons, photons, radions, and magneton packets, are more or less similar but dimensionally strange.

Quark colors would not be seen to be necessarily consistent with temperature and energy shifts. I would imagine that color shifts would occur by quantum transformation state changes. I would also conclude that under the right conditions. All fermions and bosons will be able to be affected and or transformed by electric and magnetic fields, and some may even be able to be permanently transformed to more fundamental particles and even back to a "cloud" of gravitons.

Such energy states required may only exist on Earth in the bowels of the large hadron collider.

These are only ideas which would require expert "fleshing out" by particle and other fundamental quantum physicists.

I simply must address the current theory in which photons are believed to be absorbed and reemitted by electrons in an atom. All the Feynman diagrams in the world do not prove an occurrence. A Feynman diagram is simply a pictorial representation of supposed particle interactions. Unless someone has observed electrons colliding with or emitting photons it remains unproven.

The reverse in fact may be he case. What do you observe when a laser beam intersects an electron beam. If nothing; then the current theory could be considered to have difficulties while at the same time lending support to mine.

I can hear your "victory shout" from here because of course you will cite that high energy laser beams cause Compton scattering from electrons in an electron beam. However it is fully obvious that anomalous behavior is occurring because all of the electrons and photons are not engaged by reason that the two beams don't impede each other. So my original contention still remains and your victory may well be vacuous.

My argument is dimensionally related, and my theory would actually predict the results of such collisions, because as I explained before: In high power laser beams some light is forced into the gravitos where it plasms and is then able to collide with the gravitons in an electron, so now you can have your Feynman diagram and calculate the same result either by Compton or Einstein!

This however in no way proves that photons generally collide with electrons within the atomic orbitals.

CHAPTER 8

THE GOD CODE

Now protons are all different and as well as all of them existing in the gravitos some also have extensional existence in other dimensions as well. Perhaps some quantum and or fundamental particles exist on or across the delineation regions or branes.

Protons have a proclivity for the acceptance or rejection of various frequencies of light dependant on the elementary, dimensional, atomic and molecular makeup which is subjectively dependent upon energy, electrical, magnetic, Fermi, band and motional effects at the whole atomic/molecular and macro levels.

The theory that some protons exist in the photos and gravitos, and some only in the gravitos being combined with other protonic dimensional variation is related to many observances of the behavior of light. For example it can explain how light travels through certain solid and gaseous media and not others and how it is affected by crystalline and other affects as previously outlined. Also light is more readily emitted by atomic matter in which the proton exists in the photos for obvious reasons. Also emr is more able to be emitted because of other dimensional shape shifting of protons.

The characteristics of matter apart from electric fields are theorized to be determined by protons existing or not in the following extra dimensional arrangements that are specific to atoms within matter. The characteristics are formed by a 4 bit code sequence. Atoms with protons in different dimensions will likely to be observed as having different energy states. (In "classical physics" this is thought to cause "band gaps" between the valency and conductivity orbitals, so allowing small photons of one color through and blocking large ones of a different color. How then can this effect block small ones and pass large ones if such should be observed? Also this is problematic when there is obvious confusion as to whether light is passed through an object via either the band gaps or the nucleus. Both of these are mutually exclusive, so in analysis, both create difficulties). I.e. A "wave" should pass through band gaps and bypass the nucleus. Also the band gap theory may be flawed by evaluating atoms as two dimensional realities only. Is this another absurdity in scientific modeling? See the explanation of photoelectric affect in the following chapter.

The following is a theorized proton dimensional occupation sequence which is shown to cause varying properties of matter. Some of the properties are only observed in man made materials however they do exist and may even be discovered as natural occurrences in other places in the universe.

Bit significance, (1= proton in dimension 0=proton not in dimension)

1	gravitos 2	photos 3	magnos 4	propos
1234	bit numbers, examples			
1000	transparent, non magnetic, emr transparent—glass			
1001	transparent, non magnetic emr opaque—diamond			
1010	transparent, magnetic, emr transparent—meta plastic			
1011	transparent, magnetic, emr opaque—meta material			
1100	opaque, non magnetic, emr transparent—some plastics			
1101	opaque, non magnetic, emr opaque—lead, copper . . .			
1110	opaque, magnetic, emr transparent—meta plastic			
1111	opaque, magnetic, emr opaque—steel			

This is a four bit digital code.

Once I realized what I may have discovered here, I almost fell off my chair, and I have no hesitation in calling this the GOD CODE for some of the characteristics of matter.

Now these characteristics apply mainly to elements and molecules. Complex matter such as large atoms and molecules; cellular objects and agglomerations will exhibit these characteristics in a more or less substantive manner. It also applies to the state of matter called plasma. (But not light plasma). The lighter gaseous elements seem to be prone to having their protons in the gravitos only, which increases their strong nuclear bond while larger atoms may have a dispersal of protons in different dimensions resulting in some of the observed characteristics of the element.

This dimensional shape shifting also extends to molecules such as Sio4 (glass). Another interesting observation is that elements with protons in the photos and the propos are more likely able to emit light and emr. In opaque materials not all of the protons are necessarily in the same dimensions. So the description of protons as a swarm of bees almost blocking out the Sun is apt. (as in a very thin otherwise opaque objects passing some light)

As the object becomes thicker it is similar to adding another swarm until the Sunlight is completely blocked. At this stage adding more swarms has no further effect and the object is simply opaque at a certain thickness. E.G. Gold can be beaten so thin that some light will pass through but at a certain thickness it becomes opaque.

Now having offered a simple explanation I will now tender an idea which at the same time as appearing radical may also be quite fascinating.

You are all no doubt aware of electron quantum numbers. I have wondered for some time that if the atomic force (which is extremely powerful) between the nucleus and electrons is quantasized. And considering that the average eigenvalue at all points from the centre of the nucleus to the outer orbital is a real value, and that there is a bilateral force effect between electrons and the nuclei; it stands to reason that quantum behavior of electrons must be caused by quantum behavior of nucleons. (More specifically: protons).

It would then be reasonable to suggest that nucleons have quantum numbers themselves which are energy state dependant and more importantly affected by the position of the proton in the individual nucleus. I will give protons a -ve quantum number becoming more -ve towards the centre with the outer proton layer being at zero. I would also suggest that this can be quantum "integer step" related to the strong nuclear force basement energy state, which of course has nothing to do with causing the changeability of energy quanta states in protons.

The case I can now make is that the dimensions that protons exist in can be determined by their quantum "Fermi-group" number at ground state. Quantum units apply to all Fermions.

When another atom with Fermi level connectivity becomes bound at the outer orbital then the proton quantum numbers of both atoms may change at particular quantum significance or "snap" levels and so cause a change in the properties of the material by dimensional affects.

In the future I can almost predict that an observational connection between protonic quantum theory and observed properties of materials will lead to knowledge of the connection between both proton and sub-nucleon quantum states and dimensional states.

Solid matter with protons exhibiting a force (read quantum number) guaranteeing acceptance of a particular frequency of light will absorb the light if the proton is located in the photos as is depicted in the code from the dataset above, and the material is then declared to be opaque at that frequency.

If the proton is located in a medium with similar acceptance but in this case not positioned in the photos, the photon will enter the object by the protonic attraction but it will be unable to connect with the proton now being in a different dimension, and it will pass right through the massive object,

IE glass, water, air etc. However the photons will travel a greater distance because of vector force resultants within the object acting on the photons causing observed velocity loss.

An element exhibiting both of these two characteristics in differing atomic bond arrangements is carbon, as graphite or diamond respectively. It is not the crystalline nature of the diamond that determines its transparency. (However it probably affects its refractive index and its critical angle). In that particular atomic valency arrangement, the proton has switched dimensions from the photos to the propos and the element is so able to pass light.

CHAPTER 9

LIGHT: WAVE OR PARTICLE?

I hope that by now you are in deep contemplation of the possibility of this theory of multidimensionalism (which because it seems to offer an explanation for so much in science), you might concur that it probably should be developed further, and possibly into a robust scientific theory.

Having said this, it matters little if the current theories are adhered to unless one wishes to perhaps further the micro developments of say; media event horizon evanescent effects controlled perhaps by photons themselves. Such an idea should even be more enticing than fiddling with such things as slower refractive index sensitive semiconductors. What may even be more exiting; perhaps is the possibility of photonic control of gravity? However for now let's just take a good look at light.

Wave theory is a traditionally valid method of understanding and describing light in a way that can be presented graphically and comprehensibly but it is a bit like representing fractal geometry by drawing a spiral.

First of all, I will begin with just a simple real world observationally generated question. If light was propagated by waves wouldn't we notice interference and nodal variations in everything we see? The facts happen to be the exact opposite! Everything looks crystal clear to me.

Fraunhofer and Fresnel theories and all other photonic science remains valid with this explanation of the behavior of light, because vibrating photons and strings/packets of photons can behave in strange ways, so complex photonic streams can in "general" theory be reduced to the assumption of wavelike or perhaps even spiral motion and result in the planar and spatial effects and quantities so realized, but I will be presenting a "wave" particle theory with a big difference to conventional theory. The wave does not actually exist but simply traces the vibration/rotation or spiraling etc of photons in a three dimensional way, yet further complemented by the effective behavior of the photons caused by extra dimensional affects.

This theory contends that photonic emission is fully effected by protons.

The behaviors of single and multiple photons can produce different effects as can high energy lasers, and to a more unknown degree; intense light emanating from stellar and greater universal light sources.

Before I go any further I must address a conundrum of logic that exists with current theory of how light "appears" to slow down in a transparent medium. To address this first we should take a look at how light behaves when it strikes an opaque object such as steel.

Disregarding for the moment the slight reflectiveness of steel; the best way we can analyze this is with high energy cutting lasers. It is obvious that the steel "captures" the light energy (and this leads us to the fact that photons are packets of energy!) and it would be logical to assume that the surface atoms are the first to receive and convect or re-emit the energy of the photons. This causes an immediate temperature rise and an immediate reemission of light from those atoms in completely random directions. I need go no further with describing the metal cutting process because I have achieved the main observation with pertinence to the subject of light speed attenuation in media. The point being that light entering atoms if re-emitted does so in RANDOM directions!

What then causes light entering atoms of a transparent media (as per current theory) to be passed on to the next atom without directional change? This is a valid question which should have a logical answer if the theory is correct and well understood.

However the next problem with standard theory creates a serious difficulty. I must ask: If light slows, or even appears to slow by atomic delay after entering a transparent medium, the incoming light behind it must either pile up and run into itself at the surface or pile up in the first atom/s of the media respectively. (Of course you will immediately reach for the relativity gun! If you can't explain something just change a constant in a formula. Hey that works every time! funny that!) I declare without reservation if any scientists of previous generations did that we would not have any science at all!

Because current theory assumes that light is passed from atom to atom at "c" and it is the transfer delays through atoms that supposedly cause an apparent slowing of the light, (I will deal with this problem now and the first one further on). and notwithstanding the fact that as I have already pointed out (and this has substantiation from quantum mechanics) that atoms re-emit photons instantaneously under energy state restrictions. I.E if they are in an energy state enabling them to re-emit they will do so instantaneously). So what causes the delay? Perhaps it is because atoms have to receive enough energy from photons to be able to re-emit them. After all aren't protons fermions and therefore subject to quantum units? If this were to be as supposed, then velocities in media would not be stable or constant, especially at different temperatures. Are they? (This is a difficult question because I have not stipulated any particular temperature, but you know what I mean!)

If the photons (or classical waves) at the horizon of the media have photons piling up into them then their energy state MUST increase and they will either re-emit light in random directions, or pass the energy to atoms in their proximity in a balanced manner and the media would no longer be transparent. In this case also the boundary layer should not only re-emit light it should also heat up and if a high power laser is involved, it should melt or vaporize the so called transparent object! I ask you: Does this occur? Then please explain where the photons go? The only other explanation possible apart from "band gap" theory (which I have already demonstrated to be an absurdity) is that the extra photons must jump straight around the medium at faster than "c" and recombine on the other side. The fact that my theory eminently solves this problem will soon become clear and not as pretentious.

The laws of proton-photon force effects.

1/ A photon requires a proportional force with similar but possibly reverse eigenvalues to change tines in a linear manner, from zero force at zero degrees, to force 1 at 180 degrees. This means that a force of 0.25 will cause a 45 degree tine switch in either direction depending on the circumstances described.

2/ Only a force generated by protons as "density proportional" near-field force can affect photons. What that force is (as with many other quandaries of science) I will or have addressed!

3/ Photons will attempt to switch tines in the projected direction by the force communicated by protons but with respect to the overriding law of the conservation of energy. Force vector reflection is similar to the supposed wave reflection.

4/ Photons will lose some energy when switching tines. (Whether or not, it is measurable).

5/ If a surface is atomically flat and reflective below the angle of refraction the photon will exhibit almost perfect elastic rebound.

Now it is necessary to broach the subject of the behavior of light at the event horizons between media of varying densities which are deemed to be transparent. Photons and receptive protons have a mutual attraction depending upon the constraints mentioned in the previous chapter. With respect to the maintenance of the conservation of energy; if a greater force is required to cause light to enter the medium than reflect, it will switch tines, (which indeed it does) and reflect at the same angle as the incident angle (which is the angle of the maximum conservation of energy). These actions of light are all the result of vector force and energy law.

Another contention is that reflected light attenuates slightly but does not change velocity. (The glib theory that the energy sum of reflected and refracted light equals the incident light energy doesn't make any sense within the frame of reference of the known observations in the natural universe; that every action occurring between two objects results in a utilization of, and therefore a loss of energy). This could have implications of inelastic rebound with subsequent relationship to Compton's scattering angle. (At this stage it remains unclear).

Assuming the protons in the media about to be approached by photons are receptive to those particular photons and the media is transparent then the following is the case. Photons by this theory require a force to enable them to switch tines. To reiterate: This force is assumed to be linear between a force of zero to maintain direction on a tine to a force of 1 to cause a 180dgree reversal of the photon etc. So to show the maintenance of the conservation of energy, and to reassert what I have just written: If a greater force is required to make light enter the medium at all than switch tines to reflect, it will switch tines to reflect. Refraction occurs when it takes more energy to switch tines to a greater angle to reflect, than to enter the medium. This occurs at the critical angle. This is all the result of vector forces between the photon and the protons at and near the event horizon.

This also leads one to realize that the force that causes light to bend (actually switch tines) around planets is not gravity at all, rather it is through the effect of medium through far-field "summative protonic force attraction" emanating from the massive object causing the light to switch tines! I do not have a problem with Einstein's ring only with calculations re the amount of bending.

Light becomes attenuated in two ways. 1/ The loss of vibrational and or rotational energy, which is caused by graviton depletion. This may eventually culminate in the complete depletion of the photon to non existence. 2/ loss of vibrational and rotational energy by being forced to change tines. Apparent velocity loss however does not contribute to the loss of energy of light. During refraction apparent velocity loss by this theory is caused by the change in the DISTANCE light travels in a medium.

This is not a function of time which does not change, rather the value of "c" only appears to have changed upon exit from the medium and therefore the incoming photons don't pile up. I can speculate that it is the continuance of multiple opposing vector forces upon the photon in media which causes a wavy/coiling and even a path of "near-field" wandering type propagation with a subsequent increase in the distance of travel that causes apparent "c" change. If the media material is consistent then parallel tines of photon strings will follow similar paths and exit the medium, still in parallel, within understood constraints.

Velocity reduction is not caused by a loss of energy or vibrational frequency and the light remains the same color but somewhat attenuated. The reason that the angle of refraction is proportional to the refractive index related velocity change is caused by the protonic force patterns within the media being proportional to the photonic force requirements for the photon to switch tines, which again is the angle causing the best conservation of energy. Light is also affected by a law similar to a reversing of Maxwell's field laws of increasing effect of protonic force as it approaches the media.

Photons also have a reverse "vector force affect" on atoms such that vibrating/rotating spirals of photons can cause changes in the vibration/rotation of atoms. Photons may not only vibrate and affect atoms; they may also rotate in such a manner that vectorally adjusted from their vibration they might be seen to rotate on a central axis as well. This rotation also contains "kinetic" energy, (not because of non existent angular momentum but) because of near-field, force-field dimensional affects.

Sympathetic vibration and possibly, rotational activity between high energy photons and atoms has been noticed to have some strange and amazing effects, even to the lowering the temperature of an atom to near zero k by laser emitted photons vibrating the gravitons (energy) right out of the atom and causing light to actually stop!

Now we will address the subject of reflection off a transparent media by particle theory in greater depth:

As a photon approaches the event horizon of a transparent media it becomes subject to the "protonic photaic" force field exerted by protons within the media via the namesake dimension.

Wave theory assumes a perfectly flat surface of atoms and the shiny surface simply reflects the wave at low angles of incidence by absorption and re-emission of atoms. Problems with this theory apart from the difficulties already stated may also be acerbated by analyzing several facts. 1/ the surface of a reflective medium can be far from flat and shiny at the atomic or molecular level. 2/ some reflected light has been observed to travel parallel to the surface of a medium. 3/ some light reflects and some refracts at the refractive index incidence angle. 4/ Waves appear to polarize upon reflection generally weighted in a certain orientation. This effect may be assumed to reasonably occur at or near the incidence angle of refraction but it is an enigma of wave theory that it seems to occur at all reflective angles independent of polarization which should not be the case if light simply reflects from a shiny surface. 5/ equally polished surfaces of different transparent media, have different reflectivity. I. E. light loses more or less visibility by diffusion. Even though these "wave theory" ideas are somewhat simplistic they still remain the popular view to date, and I accept that wave theory still remains scientifically useful in general theory.

This particle theory of light by vector force resultants at the atomic and quantum level explains and resolves the above problematic phenomena.

One other factor that must be mentioned is that the lower the angles of incidence at reflection the shallower the penetration of the photon into the atom (not as far as the nucleus) and so the atomic force remains proportionally as low as the angle of incidence so the angle of reflection remains the same as the angle of incidence.

Because a surface is never atomically flat their will always be some random diffusion, reflection and refraction, and at the critical angle this effect is maximized. Photons that are detected to run parallel along the surface of a medium have simply by collision with other reflecting photons been crowded on their tine and forced into the gravitos and when they finally return to a tine they will choose the tine which contains the most attraction to the protonic media. But the photons in this situation will in no way lose such energy required to change tines to enter the media and nor do they now need to reflect. This also suggests that some light "plasms" at the event horizon of media. This has particular importance when it comes to high power lasers and the behavior of light so caused, as will be explained.

Light tends to become slightly polarized at or approaching the critical angle because vertically vibrating photons vibrate deeper into the atom than planar vibrating ones and this results in some reflection as well as refraction. This is well demonstrated using polarized lasers.

Color variable transparent materials have variability of receptive protons to the color vibrational frequencies of the photons so received and the elemental or molecular material has receptive and rejective protons existing in the propos or photos as the case may be and the photons of the color vibration of accepting protons pass right through the object on the vacant tines because those particular protons are in the propos. This is what causes visibly colored media. Various elemental impurities in the media will cause band and line spectra distortions.

Light with a ninety degree angle of incidence is the easiest to understand as the forces on every side as it approaches the horizon have a net result of zero and the attractive force of combined receptive protons (existing in another dimension) is of insufficient force required to make light change tines to 180 degrees, and the eigenvector/value of forces remains unchanged. (IE not a mirror)

This photonic entry into a media is a thoroughly complex situation with gravitons being emitted and absorbed and reabsorbed by protons and photons alike by the previously described interactions of particles and dimensions. The photon is seen to lose some amount of energy and therefore luminescence and because of this fact, no object is completely transparent. Imperfect materials also cause aberrations.

Once a photon refracts it "discovers" that the forces within the new media are proportionally different than the forces which were being exerted upon it in the previous media. The photon can no longer reflect to conserve energy. It has seemingly been tricked into entering the media, and if the new media is of greater density or differing molecular configurations it experiences larger vibrating and conflicting forces acting upon it, which causes it to switch tines an innumerable number of times. As we saw in a previous paragraph the distance it now travels through the media has now become longer and it only appears to have slowed down. There is therefore no Doppler shift change in the frequency and hence color of the light.

This Doppler problem is only answerable in classical theory by the idea that light can change its frequency and not its speed and the converse would be: (although unacceptable) If it were able to change its speed it would not be observed to experience a frequency change, both ideas are relativistic and absurd and would have to occur without any other known cause! Prior to this particle theory this has been a "black hole" in science and conveniently ignored to a great extent and simply glossed over with the relativity bullet.

Let's face it relativity is so conveniently able to be used as a "band aid" on many occasions, but considering the frequent necessity for its use, and the fact that on many occasions the wounds are simply too large, The patient if clearly in view, is seen to be bandaged up all over yet hemorrhaging and close to death. "clear"!

Once light has "bent" or entered the perfect media there is nothing in its way, because as it passes through atoms all the fermions are on different dimensions than the photons and the light gets a free ride though the media and still at "c".

It must be noted that the photons are also existing in the force field dimension and are therefore subject to the forces of the protons that they pass on the way, causing them to react accordingly, and the photons now behave more like a wave (or to put it simplistically a squiggle), and travel in a wiggly line. The wandering path is controlled by the density and molecular or atomic and crystalline positions of the protons, some of which can cause the photons such a convoluted (and even very contorted paths are possible) that the observed slowing of the speed of light through some media is very great indeed*. I will reiterate; that this action causes the photons (while still traveling at "c" and continuing their own vibration signature) to appear to have slowed. (Some media appears to slow light to incredibly low speeds which are almost beyond comprehension. I believe however that light plasma effects and not path convolution is the cause of this. However this still remains unclear. *This is also co-dependent on the energy level of the photon. A higher energy photon such as an x-ray will bend less and be more likely to pass through a media which is mildly translucent or even a media that may be opaque to normal light.

If protonic attraction causes light to bend around planet sized objects, won't it still "bend" as it passes by very close to a smaller object? The answer (by theory) must be yes, but such an effect has only a remote possibility of being measurable against extremely large and very dense objects on Earth. In proportional comparison with planetary sized bodies this bending may only occur within one or two atom distance from a real world sized object and the tine shifting would only be small and probably masked by the random atomic effects at the surface of the object. It is theorized that the closest photons to said atoms (determined by orientation of oscillation i.e. polarization) would be captured by the object. This may also explain why the shadow of very small object exhibits a dot of light at the centre.

The situation for light running parallel and very close to a transparent media may be the same but not necessarily. Similarly; for a mirror which is made from material with unreceptive protons. It would depend on the thickness of the mirror surface material and density of the whole medium.

This might sound problematic if you assume that the surface of the transparent material "actually" becomes a mirror at low angles of incidence. However according to the reasons that light enters or reflects according to this theory one must remember that it is protonic forces and not the reflectivity of the media horizon that causes this.

There maybe a situation at very low angles of incidence (IE the grazing angle) with transparent media that some photons will enter the media and exit via another proton horizon in a random fashion. This may not actually be observable unless one has a laser of only one wavelength and a single photon stream. This effect may be technologically significant!

Electrical manipulation of media has already enabled technological uses such as LCDs and even though this is not manipulation at the atomic level, it is still interesting to suspect that future technologies becoming available with atomic manipulation could be many and varied, with some even novel. (Invisibility! anyone?) What would you see if light striking a transparent object was totally refracted or reflected or diffused by atomic control and you could modulate the subjects in many different ways purely by photonic or electronic means? Currently magnetic and electric fields have no affect on the light because photons have no charge or dipole. Light is not electromagnetic period! However this may change in extreme environments. i.e. in Magnetars.

Creating extreme environments allow science to remove matter from the influence of the Eos and this constitutes a method of dimensional manipulation. Other means should be sought.

If we could find a way of popping protons in and out of dimensions at will. Imagine the possibilities! (Refer to chapter 7)

This theory of the particle propagation of light, by addressing and answering difficulties with standard models helps support the multi-dimensional theory per se.

The reason that white light diffracts is because photons of differing vibrational frequencies by possessing different energy components have differing force requirements to switch tines, and so they diffract at an altered angle according to changed vector force resultants.

Prisms I will not address because it is just more of the same resulting in further refraction, but in so saying I still must address refraction from a dense to a less dense medium, total internal reflection, mirror reflection against absorption, refraction, double diffraction and diffraction gratings. Also to be addressed are Fresnel lenses, lasers, plasmas in more detail, and spatial displacement of "laser" light at total internal reflective event horizons in some materials.

Refraction from a dense to a less dense medium is simply the reverse vector force analysis compared to the opposite direction. Total internal refection is still vector force related to photons and protons.

Total internal reflection can be treated in the same manner except for intrinsic behaviors which I will shortly address.

Now to the mirror: A mirror contains (contrary to what you may think) super-receptive atoms to several layers of atoms thick. This atomic density is sufficient to present such a force to the proton that it is sufficient to cause it to change tines up to 180 degrees and even though the medium is receptive, the protons are concurrently in the photos and hence are able to exert direct and extreme attractive force to the photon at the horizon atom. This force is at a level which is able to cause the photon to switch tines to 180 degrees and this in effect tricks it into reversing direction because it "thinks" that in so doing it will conserve energy. This explains the effect of polishing normally absorbent metals to enable them to reflect, because in that case an almost perfectly arrayed force of protons is presented such that it will then be atomically reflective, albeit with color anomalies as previously explained.

Spatial shifting of laser light during total internal reflection in some materials would be caused by the high angle of reflection and tine/dimension shifts caused by photon-photon collisions causing interactive vibrations resulting in interference phasing (or plasing) which in turn cause the photons to shift tines and effectively bounce along the internal surface of the medium in an evanescent wave until they escape the overload interference of the incoming laser light and are now able to reflect as per the normal explanations. I must admit that this is just as hard to describe as it is to understand. Put simply the high energy and density photons reflect and the incoming photons collide with the reflected photons and "plasm" in a vibrational sense because after plasming they become "tagged" by incoming photons and forced to the angular tine traveling back to the event horizon and by so doing they or other photons reflect again, and the whole process is repeated until the photons are out of the collision zone and can reflect unchallenged.

Diffraction gratings act according to this theory as a series of lines of "size dimensions" necessary to create minute atomic/molecular level prisms with respect to the size of vibration amplitude and frequency of the photons.

As I have indicated before; light plasma is theorized to consist of photons, either individually or en masse, existing (and even able to be stationary) in the gravitos. Plasmas are observed in nature traveling at relatively slow speeds, such as finger lightening balls as well as ground proximity ball lightening just drifting around. This plasma may consist of light plasma, or in other terms cold plasma. (This is because hot plasma would lose sustaining temperature too quickly to exist in the atmosphere for many seconds as is often observed. E.g. the plasma caused by lightening is usually gone in milliseconds).

Apart from fusion energy, other plasma research that may be carried out could entail content based on the following . . . Luminescent light plasma may be able to be manipulated to cause it to soak up gravitons and not re-emit them and in so doing cause anti-gravity effects, which could lead to anti-gravity drives and shields as well as artificial gravity. It may become possible for hot plasma to be utilized as an impact shield. Containment and protection for a starship utilizing such faster than light, gravity drive technology would be an absolute necessity.

The occupants of a ship in similar technologically designed anti gravity suits (or an internal pod in the ship surrounded by such) would not even be subject to the laws of motion because they would be effectively massless. Atomic, molecular and cellular bonds would not be affected. However an artificial gravity may be able to be induced by high intensity laser collision: (Just a thought!) Protection of the occupants from the light/heat plasmas could utilize heat tiles and artificial sapphire and fused silica mirrors. Such a starship may appear like a small Sun approaching and scare the living bleep out of any alien observers of distant worlds. (I digress with theory based fantasy).

Einstein's slit and all polarizing slits act as they do by force interactions between photons and protons in the vicinity of the slit material. The movement of atoms being protonically attracted to the photons passing through the slit will also distort the slit, such effects should be taken into account when analyzing data acquired via the slit or holes. All light energy loss can be accounted for. Some photons are absorbed by the atoms near the slit either by direct atomic absorption or plasma affects.

Interference patterns produced through light experiments do not prove wave theory. They only prove a patterned dispersal of photon densities caused by the transverse vibrational vector force resultants at the atomic event horizon of the hole or slit so used. This is because of elasticity and vibration of the holes/slits caused by accelerative and deccelerative vector forces acting on both photons and atoms at the quantum level. Any supposition that the slit or holes are stationary objects (Meaning the matter in near-field vicinity) when light passes through will lead to incorrect conclusions. So for instance if two photons are traveling side by side and one passes through the centre of the slit and one passes to one side, one photon may switch tines in responses to the instantaneous quantum force applied, and it will "bend" in a slightly different direction. Photons of different colors will diffract differently of course.

Light has been shown to change tines when two high power laser beams collide and an interference pattern is noticed at the point of collision (in a vacuum!) This shows that because photons don't bounce of each other (they are supposed to be more boson than fermion), so they plasm in the vacuum (not being a true vacuum according to this theory and switch tines that send some of them to your eyeballs so you can see them. Otherwise there is no other theory that I know of that can explain how "bosons" colliding in a vacuum can redirect light! It is most likely that dark areas observed in the pattern are actually a true vacuum; devoid of any matter at all.

Lasing is probably caused by propos protonic effects causing the photon to plasm until the energy level required to force them to jump onto a tine in a particular direction (so caused) is reached and they then "lase" by choosing parallel tines (so caused by the same quantum energy level being reached by all photons at the same instant). (Refer to interlocution).

This postulation need not be especially accurate science, and the standard theories of lasing may well be more accurate. This could be ascertained by experiments into the ability for high energy light collisions to plasm and create and absorb gravitons enabling them to create or delete gravity. This suggests that the ultimate vacuum is possible. (See I have proven that "nothing is possible!" Ha ha). Seriously; If what is created is truly the ultimate vacuum then according to this theory it is really a "hole in gravity" the importance of which might be unimaginable and extremely significant to future science and technologies.

Highly reflective objects such as mirrors have protons that reject photons and the force required to switch tines to 180 degrees is less than the force required to make the photon enter the object. This is still by the photon assessing which action will result in the least loss of energy and as usual it maintains eigenvector parity.

This all makes a photon seem intelligent doesn't It.? Not really the photon simply has internal ability to feel the rejective or attractive force of the protons at the event horizon via the force-field dimension. All the photon gets to know is if the proton has got enough attractive or rejective force to cause the photon to exhibit certain behaviors dependant on its own attractive force? That is all the photon can react to. Is it programmed to do this?! I think not! The forces are more likely to be caused by something as mundane as sympathetic vibrations, than anything so profound.

An explanation of the particle theory of the photoelectric effect: It is thought by particle theory to be mainly caused by incident photons carrying quantasized "data" (per Plank) which has a rebound effect on electrons, or indeed a "photo-photonic" affect if the incident atom is in a full quantum state and the Fermi state is right etc.

This is very similar to the method of reception of ramaton particles in the radio reception explanation in chapter 10. If the material being incidented has receptive protons in the photos then the higher vibration frequency (and perhaps even the larger size??) of ultra violet or blue photons will cause the effect more than red photons which don't actually cause the effect to occur at all.

CHAPTER 10

ELECTROMAGNETISM

An electromagnetic field is one of the fundamental forces of nature, so with regard to magnetic and electrical theory there is no change to classical laws and formulae.

If as previously described an electron has an imbalanced—ve dipolar electric charge at ninety degrees to its magnetic dipole and does not have planetary motion around the nucleus, then the following is likely to be a valid explanation of electromagnetism.

Note: The jury is still out on causality or causation of charged particle angular momentum and the relationship between spin moment and magnetic and charge fields. This appears to be very problematic and there is not even any consensus whether an electron is a non existent point source, or whether it indeed has continuum charge/dipole dimensions and shape. (I agree with the latter because I contend that all matter must have size even if it can never be visible by human observers).

If this is not the case then electrons must simply be interference crossing points of the vibrating charge and magnetic fields. This is disproved as a postulation because electron beams can be forced by applied energy to leave the atom and atomic matter and be attracted in a beam of particles to a high positive charge in a vacuum. i.e. The old fashioned vacuum tube TV set. This also proves that an electron is both a dipole and a negative charge particle.

Something else of interest is that there is no consensus in science regarding any equation for the motion of charged particles.

DESCRIPTION OF E.M. MECHANICS: The explanation by this theory of current in a conductor causing the induction of a magnetic field:

When a conductor of resistance "r" has an emf applied the electron flows away from the negative terminal of the emf source because electrons have a negative charge and are attracted to the positive terminal according to the basic formula e=ir. This is classical physics.

What is not often understood is that the emf causes a near-field electric charge evenly applied to each atom in the conductor according to the formula (atomic charge) Ac=e/n where "e" is emf and "n" is the effective number of series atoms in the conductor. This causes a change at "c" to every individual atom's normally concentric field charge, which causes it to become an elongated, lopsided field charge. It is elongated (mainly affecting the outer conductance band) and directional in the "length" direction of the conductor.

This changes the orbital nodes around the nucleus and weakens the charge field strength of the nuclei equally and randomly at right angles across the conductor, which allows electrons in the conductance band to wander more readily between atoms and be able to leave the outer shell (in the manner to be described), and travel to the next atom in what I will call "carrier gaps" This a gap between the conduction bands of atoms which are caused by the weakened and emf charged depleted and narrowed nodal zones between the atoms at right angles to the conductor.

This non uniform nodal elongation causes the electrons to "crowd" the nucleus to keep their charge and magnetic dipole in the orientation determined by the (averaged) nuclear dipole.

This lopsided elongation of the charge field around the nucleus enables the bunch of electrons crowding one side of the nucleus to force its nucleons to change orientation from their normal 45 degree magnetic balance state to an orientation at or about 90 degrees to the angle of charge field elongation. This reorientation of all the nuclear dipoles is what immediately gives the conductor the behavior of a magnet and it exhibits say a N/S orientation at right angles to the conductor. This is the same from the centre of the conductor all around it towards the outside. If the emf is reversed then the magnetic orientation is reversed albeit with real time delay.

Now electrons in average nodal orientations around the nuclei provide the nuclei with zero net magnetic change, and they only exhibit a change in their orientation, and the electrons by self repulsion are easily able to hop from the high negative region of one atom to the positive region of the next because of the applied potential difference between atoms but subject to Pauli. However because of ultra high frequency nucleon vibrations we expect to see electrons hopping in individualized pulses through the carrier region between atoms, the electrons now also have an effect of strengthening the overall magnetic field produced by the conductor in the following manner.

As an electron passes between the now transversely aligned atomic magnetic dipole, the strength of its own dipole adds to the atomic dipole in proportion to the number of (sympathetic at "c") pulsed hopping of electrons traveling between atoms at any given point in time.

We now have a pulsing magnetic field that has field strength in proportion to both emf and electron flow initially and then perhaps solely a function of increasing electron flow until saturation (This might help explain hysteresis). The proportionality of the field to atomic dipole movement will not be determined (as that is another issue), but neither can its existence go unrealized, because both atomic dipole effect and electron flow are both a function of emf until the nuclear dipoles are firmly at right angles, (Perhaps when that occurs it could be likened to a reverse Paschenback effect)* and both are similarly proportional. Further research is required.

*This would only be expected to occur at superconducting temperatures, where the quantum number possibilities are vastly increased. (Because the Eos is dysfunctional at such low temperatures perhaps the Pauli principle could even be overcome??) The inner orbitals would be hugging the nucleus. The conductance band would have likely combined in a unity web with every other atomic outer band. The Fermi level would be invalid. The band gap would be enormous and all electrons in the outer orbital would be fully engaged but only on the surface of the super-conductor because without the Eos the magnos is ineffective. This would result in extremely low electrical resistance and an extreme magnetic field when current is flowing. When no current is flowing the conductor would be resistant to magnetic field lines of force because of the rejective force (previously described) because of the same dimensional reasons.

In this state the outer band has been dislocated from the nucleon Hilbert set and the vibration of the atoms is very slow. Another supposition is that the vibrating nuclear electric charge would be severely weakened by the low energy state and the electrons would no longer be affected by having to "hop" and they would also cease wandering and simply go with the "flow" along the "skin" of the super-conductor. Now back to STP considerations.

The pulsation frequency of the magnetic field lines now expanding outwards in concentric lines of force, (more like tubes of magnetic force which are parallel to the conductor) are a function of nuclear vibrations. This is the only time we see a field that apparently has no poles. It does actually: The poles of the electromagnetic conductor are the ends of the conductor! The overall field would be expected to be vibrating except for the averaging affect of trillions of atoms and somewhat fewer electrons which causes a steady state magnetic field consistent with a constant emf being applied.

THE CONVERSE EXPLANATION OF A MAGNETIC FIELD CAUSING A CURRENT IN A CONDUCTOR:

Now we will address what happens in the reverse case of magnetic lines of force cutting through a conductor at right angles, and we will see why the electric effect is in inverse proportion (not necessarily linear) to the reduction of this angle.

Magnetic Lines of force actually exist* and keep distance between each other by repulsion due to the standard explanation in classical physics. The density of the lines is not a necessary factor in this example. *This can be noticed by the flattening of an electron beam in a magnetic field.

When a magnetic line of force cuts though a conductor, it applies a force on both nuclear and electron dipoles at the same time as it approaches and with a squared effect proportional to closing distance.

We already understand that electrons have a magnetic dipole and an electric charge at right angles to it and it will always attempt to keep its charge somewhat oriented to the attractive charge emanating radially from the nucleus and with its dipole oriented to the nuclear magnetic field within the constraints of the four forces which are affecting them.

Along comes a line of force which passes through the atom and the dipole of the atomic nucleus moves into alignment with it, and the electrons also come in to alignment. The only way they can do so and still keep their charge alignment tied to the near-field nuclear charge is to move to one side of the atom with their dipoles more or less oriented with the direction of the nuclear lines of force. In any other position around the atom they cannot retain both dipolic and charge alignment at the same time. So they group close together on one side where than can maintain their charge pole at right angles to the magnetic line of force, (Ignoring diamagnetism for the moment!)

This then causes a charge field displacement in the whole atom because by reason of the grouping of negatively charged electrons to one side, it is now more negative on one side than the other, and this causes the formation of the atom into a tiny electric charge cell*, and because the length of the conductor is at right angles to the 90 degree cutting angle of the magnetic force line the charge is oriented so that if the magnetic field line is north at the top and the line of force is moving away from you into the conductor, the electrons will gravitate towards the left. Conventional flow will then be towards the right. This follows Fleming's right hand rule for generators.

This is caused by the way the nuclear charge and magnetic dipoles become oriented and it is not a function of the electron which is able to move to any side of the nucleus. This can only occur if the quarks in the proton are in a planar arrangement such that the nucleonic magnetic dipole is also planar and is at right angles to the line of dissection of the two 2/3 positively charged up quarks. This causes the individual protons of each atom to affect the nuclear charge which then becomes lopsided, which in turn forces the electrons to group near the more positive side.

The conduction band is unlikely to be affected because of reasons being already described herein. This quark/dipole arrangement is only found in objects with protons in the magnos.

We are beginning to get an idea of how dimensional forces may come into play inside nucleons by the shape shifting of quarks and magnetic dipoles! The question beginning to arise in my mind is. Do dimensions contain different and invisible "force frameworks" which align the fundamental particles? If invisible lines and waves exist then this is the next logical step.

Now we will rejoin the narrative. No CPT symmetry here! Magnetic dipole disparity is a function of both the magnetic dipole and the force line strength. This theory then assumes vibration or pulsation as spin moment and allows vertical and horizontal disparity between planar electric and magnetic dipolic moments.

The more lines cutting the conductor, the more charge differential or (potential) you have along the wire by series voltage addition which we can express with the formula* emf= Ac x n (where Ac is atomic charge differential and n is the number of atoms) and if you close the circuit while lines of force are still cutting the conductor, electron flow will occur. * This is a simplistic formula which only takes into account series charge summation. If parallel charges were taken into consideration the formula would be more complex but the end result would be similar because the parallel charges would cancel out by vector math.

If you tilt the conductor in the field you reduce the vector force direction of the atomic poles and reduce the overall potential difference. If you increase the magnetic field strength you have more and stronger lines of force cutting the conductor and more electrons will jam closer together on one side of the atom and it will also occur on more atoms at once thus increasing the charge potential and therefore the overall emf, and a greater current will flow when the circuit is closed.

This will cause either BBR to be emitted or photons as heat and even light. If we pass an alternating magnetic field through the conductor we will have an alternating current, and in this case more gravitons and even ramatons can be released causing some spurious field charge radiation (or emr).

In both of these examples the internal arrangement of force particles such as quarks, magnetons and gravitons has been changed and similar to the effect of changes caused by chemical action during bonding, energy is released from the nucleons and it makes its way to the surface of the conductor where it is emitted. In some cases where a vacuum can prevent the oxidization of the conductor, this energy emission can be seen as light.

CHAPTER 11

ELECTROMAGNETIC RADIATION: (radio)

I have analyzed the probabilities of radio frequency wave emr being propagated through space and have arrived at the following conclusions which will be explained in depth.

The first conclusion is; that even if space is determined to have low impedance to emr, there is no known medium which can be described to allow such an action to occur, and considering the fact that an "aether" has been disproved, such propagation cannot occur!

My theory could possibly provide a medium to allow compressive graviton propagation of wave fronts. However as with sound through air such propagation attenuates at a greater rate than inverse square law dictates because of energy losses to the medium, so universal observations also rule that out as a possibility.

The fact that gravitons are in the gravitos (being a completely different dimension) also provides a significant difficulty, and even if it was able to be the propagation medium, gravitons will present drag to other particles (namely each other) and there would be a necessary loss of energy in the emr wave. This would result in distance inverse proportional impedance and the emr would attenuate to the degree that radio transmission through the vast distance of the universe would be impossible.

So it has become necessary to search out the behavior of atoms and electrons during frequency voltage oscillations within a conductor to find another solution.

I will now attempt to show that electromagnetic radiation (across the "aether") at radio frequencies cannot occur as explained by classical physics.

Electric fields and magnetic fields are "near or proximity field" phenomena and although they expand at "c" (which is no surprise). they have a decay rate dependant on inverse square law.

Apart from Lorentzian relativistic formulas which are the only other way out of the enigma, and by envisaging the ridiculous idea that free space actually has no dimension. (Hey "The Emperor has no clothes!") We should take note perhaps of the fact that the permittivity (electrical) in free space is 10e-12 farads per meter, and the (magnetic) permeability is 10e-6 henrys per meter. However it is significant that even this permeability is only allowable through a rolled sheet of copper, and I certainly didn't notice one of those in free space as I drove by!

So we are directed to appreciate that the law of increasing the square of the medium per meter, is negated by inverse square law, but in doing that one must assume that free space is a conductor. WHICH IT DEFINITELY IS NOT! (Go on show us: Where is the experimental evidence that free space is a conductor?) The derivation of Lorentz's formula and others should not be by assumption, rather by working backwards from observed reality and finding out whether relativity or adding "c" into the mix; works or not.

This assumption may seem to be logical to you, but how can it if it is simply because a dimension somehow seems to have proportionality reliance to "c"? In fact it is far more likely that "c" is reliant upon dimensions as previously described. Note: the close reliance of "c" on the Eos will be analyzed in a following chapter.

What occurs in such a case as this; is that one can "assume" proof of an effect even when in actual fact the effect may be actually being caused by an unrecognized dimensional combining phenomena. This is problematic for relativistic interpretations, if dimensionalism doesn't require space to be a conductor or for the requirement of a tube of copper for emr transmission. This little unknown fact has gone unrealized in the historical formulations.

I KNOW the formulae work. It is the misunderstanding of the mechanics of the whole "emr" propagation which forces science to ride roughshod over blatant illogical dilemmas and engage in the art of "convenient exclusions". Note: because of the energy formula, relativistic forms are valid for the reasons of simplicity and there is no real need to change them in practice. Simply stated it can be expressed that it should be recognized that relativity is only a tool for calculation because like a "line" it doesn't actually exist and it is not "time" that is varying to enable such unscientific reasoning.

Even though further derivation of formulae shows that free space has real and calculable impedance by these relativistic functions (which is calculated in ohms), it is somehow incomprehensibly declared by physicists to actually not have any "real" resistance by some inexplicable and perhaps convenient lapse of logic!

A far-field Z of 377 ohms in space seems impossible to correlate with the concrete requirements for realistic application of permitivity and permeability functions? In my theory however this is simply recognized to be the impedance of the "emr conductor" being stated as the propos. There is no logical dilemma here because one cannot measure the propos and similar to the photos it can be seen to be synonymous to a theoretical conveyer belt. So the same requirements and inherent difficulty don't exist by this theory and the normal formulae can still be applied. NB But now however the dilemma has vanished.

Without relativity such problems of logic make these electromagnetic forces surprising candidates to transmit anything at all over realistic distances. Even Coulomb's electrostatic law should confirm suspicions that classical emr would be a non starter!

The fact that the magnetic affect of the Sun reaches across the solar system can be explained by two reasons. 1. The Sun has about twice the magnetic force (gauss) of the Earth, which being multiplied by a far greater volume calculates by power law into a far more size massive magnetic field. i.e. It is phenomenal! 2. The magnetic influence of the Sun which is being exerted on the nether regions of the solar system is by the emission of particles with magnetic dipoles and not directly by its field!

The actual magnetic value depreciation of the Sun's magnetic field is still according to inverse square law, so in comparison to all magnetic fields it still weakens at the same rate as other magnetic fields.

Electromagnet radiation is not emitted simply because of internal energy gain of atoms. This would result in other types of energy emission which we have already discussed. It is a radiation effect caused by oscillatory charge effects in nuclei and electrons moving in a conductor, also in an oscillationary manner. i.e. (By the "back and forth" motion in a conductor which is caused by electrons "hopping" between the outer conductance band of atoms). The atom then uses the energy gained from electron motion through syncopation to resultant nucleon motion. All nucleon energy behavior described here is determined to be quantasized and subject to Pauli and Fermi, except for electric and magnetic fields. The quantum actions are deemed to be energy related.

We can now state that this electron motion causes the atomic nucleus to vibrate in synch, because, as we considered before: the electron has a bilateral force relationship with the nucleus by electromagnetic interaction. This vibrational motion (in quantum steps) of the nucleus must be recognized as *absorption of energy from the electron shell and it requires the release of that energy, only to again be followed by the regaining of energy, all in synch with the momentary loss and gain of electrons in its conductance band orbital as they travel back and forth between atoms. *This is because whenever an atom loses an electron. The nucleus must change energy state by one quantum number. This all occurs with some elasticity and therefore proportional energy loss. The efficiency coefficient is less than one. This loss is required for the instantaneous (really at "c") "conservation of energy" in the atom and it is not to be envisaged as an overall nucleon electromagnetic value gain or loss. The energy put into electron motion is provided by the external oscillationary voltage source acting on the conductor.

Now the energy loss has to go somewhere! This electron induced force field vibrational effect on the nucleus does not cause emr, it causes the proton which is existing in the "propos" as well as the "magnos and force-field" to act in the observance to the law of the conservation of energy, (State) Which causes it to pulse out stream after stream of gravitons and magnetons*, which I will call ramatons (not to be confused with gravi-photons) at the same frequency (not necessarily phase) as the oscillator and further controlled by the electron motion into adjacent and subsequent nuclei via the propos until the event horizon atom emits the gravitons into the propos which according to the pertinent law, it causes to propagate at "c". * The ramatons trace an electromagnetic "wave" as they travel.

There is a lot going on in the atoms that needs to be addressed. The oscillatory affect of the external emf on the nucleus, does change the proton's charge but in normal radio transmission (never to negative as this would result in ionization of the conductor and it would necessarily, melt, fly apart and fail).

As we saw in the static example of electromagnetism, the electron orbitals elongate in synch with the emf and the dipoles wobble, and the electrons travel from one side of the nucleus to the other. This electron motion is what induces the proton to release radions and magnetons as ramatons to the propos all around. This occurs when the near-field induced magnetic field changes, (caused by the electron motion) interact with the nucleon dipoles causing them to emit a string of magnetons during one "half cycle" and a string of radions during the next. This is determined per atom according to its own dipolic orientation. (Dipoles in non-magnets always switch back to their original relative position when the external or near magnetic field is removed). This causes an elastic force interaction which results in electron lag which also means that the nucleus is experiencing an oscillatory energy change within it. This is because at each half cycle the conductor switches magnetic pole orientation from the centre or inside out. (Dependant on whether the conductor is solid or tubular respectively).

Now because the normal angular orientation of the nuclear poles in a non magnetized conductor is at or about 45 degrees in equilibrium (at ground state), half the dipoles in vector summation have to shift beyond 90 degrees during one "half cycle" and less during the other half cycle in the opposite direction. This means that because the conductor has impedance, the electron lag is increased and can be even "tuned" to a particular resonance by standard methods to cause ramaton "wave" reinforcement.

So in one "half cycle" any atom will emit gravitons and on the other half cycle magnetons because of differing nuclear energy disparities with electrons. (Its "next door neighbor" might be reacting in the opposite manner). These are passed directly through electron orbitals on a different dimension called the propos.

These "packetized streams" arrive at the surface at slightly different times and they overlap but on average reinforce. This requires the antenna to be of certain thickness, or tubular etc for efficiency.

Electron flow does not need to occur and INDEED IT MAY NOT especially at high frequencies*. The energy source for the transmission of ramatons is transferred directly by "bucket brigade" from the emf source as positive flow dynamics which is graviton transfer. *The antenna can be a tuned whip etc from which no electrons are emitted. There goes another dilemma. The energy loss is countered by provision at more or less a steady average rate from the emf source.

So then at the surface (event horizon) of the conductor the ramatons are emitted in a tuned stream of magnetons followed by a stream of magnetons, each related to one half cycle which makes up one cycle "wavelength" at the oscillator frequency.

These streams of particles (even though still contained within their own dimensions) travel as separate distinct packets* in the propos" and are emitted from the emitting protons at "c". External energy inputs (to atoms) that do not have a charge component will only cause BBR, photon emission and or convection. *Once "packetized"; bosons remain so, and keep equal distance from each other and more so from other types of boson packets I the same dimension, (and this will give an upper maximum possible frequency of radio transmission). So they line up and exit the conductor in "marching abreast" fashion.

This is one of the few times that gravitons are not confined to the gravitos, and as is the case with photos and Eos emissions it is concluded that their will be no measurable effect on gravity or mass during this action within the conductor. Simultaneous emissions are also occurring all along the conductor, but for reasonably effective emr to occur the effect is magnified by well known means which I don't have to get into because it is all simply theory as per standard physics. What is not standard is the fact that I contend that it is not a "combining of electric and magnetic fields" that is occurring to propagate the energy. It is a field wave of gravitons and magnetons that have no charge or mass and require no classical media for their propagation.

I will address the obvious objection. "Your propagation will still be attenuated by inverse square law as well". I'm glad you said that because that proves that you have a problem with classical theory which cannot explain how impedance in space is a constant. With this theory I do have an answer. The propos provides this impedance as part of its properties.

The objection to this will be. "You have just shifted the answer to another unknown". Exactly correct! But science is not about "getting it right first time" but if you can get science closer to the true nature of things you can enjoy better PREDICTABILITY of experiments, and two cases in point. Are. 1/ The Large Hadron Collider will not create black holes according to this theory. If they create any black "spots" at all they will simply be bubbles of perfect vacuum which wouldn't hang around for very long!* 2/ Nuclear fusion will not be able to produce a overall positive energy result, also according to this theory. LET THIS THEORY STAND OR FALL ON THESE TWO PREDICTIONS!) *Sun spots are likely to be similar.

Further evidence of this theory may be offered in that an American company has developed a wireless visibility system called RuBee* that does not transmit an electromagnetic "wave". It only transmits magnetic "waves" streams of magnetons. *RuBee is a trademark of Visible Assets, Inc.

Electrons are also not transmitted into the "ether" their job is to simply "suck energy" from the emf source and transfer it to protons by converting it into motion, and the rest is as described.

Another theory is that electric and magnetic field waves propagate through space without requiring a medium at all. i.e. I'm referring to the "ether" theory. Apart from the other difficulties that I outline herein; please explain how this can possibly occur! (Near-field charges and magnetic fields are always an inverse square law depleting force field which does not propagate independent of the point/s cause/s of the fields). Space has to be an electromagnetic conductor by some description! So far the conduction medium has not been found.

Scientists; by radio and antenna technology are able to manipulate the ramaton density pulses (radio waves) to remain in phase and become amplified and even spatially and orientationally directed.

ANTENNAE AND RADIO RECEPTION:

This brings us to the point where we must analyze why some materials such as antenna dishes reflect ramatons and some devices such as a receiving antenna connected to a coil of wire (even of non magnetic copper for instance) absorb the ramatons and reconvert them into electron flow pulses.

First and foremost we must consider the propagation media, and then the reflective or absorbent media parameters after that. So, we know from previous analysis that ramaton density pulses can propagate through any media in which the proton does not exist in the propos or magnos, which is definitely true of space, and to a lesser extent air.

These media event horizons have everything to do with reflection or reception of ramatons. Most metals according to the God code are opaque to emr or ramaton particles because their protons are in the propos. This means that they will not allow the passage of ramatons. But they are not very receptive to them either because protons do not want to receive energy of any kind unless it is in synch with their natural vibration (and other previously described restrictions), which is very high; perhaps at light frequencies and above depending on the material.

Because of these reasons antennae actually reflect/refract and generally re-concentrate the pulses in the normally accepted manner but the ramatons are received in a coil of emf energized wire or conductor. (Crystal sets work without any power because of anomalies in all materials which will cause some natural acceptance of the ramaton pulses to nuclei which causes miniscule voltage oscillations).

The emf is once again an applied force which is again moving electrons, but in this case the electrons are moving in a single direction. The proton however is again affected by the force of electrons hopping outer orbitals. Because of electron propagation delay in the conductor this hopping occurs as finite hops of electrons at a certain frequency, and by manipulating the receiving coil scientists have worked out how to match the frequency of the sympathetic proton energy state vibrations caused by the reception of ramatons, and which are no longer bi directional. So in this case the proton won't re-emit any energy via ramatons. It will simply react in an elastic quantum level electromagnetic manner with the motion of electrons hopping in and out of its outer orbital.

At the very moment in which the proton is in a state of energy depletion being at the same moment that an electron hops out of its outer orbital (it will be more than instantaneously receptive to an incoming ramaton or "two". (i.e. radio energy). This happens to all atoms in the "tuned" circuit simultaneously and the bulk of the ramatons are received into the protons.

This results in the proton now having a net energy increase as soon as another electron hops into its outer orbital, which it immediately passes down the proton "bucket brigade", which when full throughout the conductor, (Because all of the atoms are acting in a similar fashion which happens almost instantaneously, but not quite equally through out the cross section of the conductor, resulting in some undesirable effects i.e. skin effect), they also all pass some excess energy to the outer orbitals in synch with the unidirectional hopping of the electrons, and the electrons become forced by the energy cycles of the proton to decrease the electron hopping density and then increase it over and over again. (What is happening here is transference of protonic energy in a pulsating manner to interrupt or support the ebb and flow of electrons as the case may be), which when further manipulated with say an electronic low impedance "sink" such as a capacitor, it can be "tuned" and so may cause them to travel back and forth (oscillate) through the outer shells by well known electronic affects.

This movement of electrons is now either a pulsing or by LC tuning an alternating electric current which is perfectly in synch with the incoming ramaton pulse frequency by near-field electromagnetic effects (so selected by the tuning) that all the information on the "waves" is amplified and can then be more readily utilized. A poorly tuned antenna or out of tune receiving coil will cause ramatons to be reflected. Some ramatons will exhibit anomalous near-field effects because of the "nothing's perfect" law of nature.

The overall depletion or gain of ramatons is normally under control of the Eos dimension, which has been stated to be able to be overridden by local emf's*, which is the case in this example. When the emf is switched off, nucleon charge balance maintained by nucleonic graviton density is instantaneously carried out over the Eos dimension. (Other fermions and bosons are theorized to be the decision makers regarding ramaton reception, inter-nuclear transference, and transmission). *actual electromagnetic transfer in atomic media occurs at about "c" when caused by local forces in the force-field dimension. Background radiation in space is most probably the most noticeable effect of the Eos, and charged particle propagation is also a balance being promoted over time.

This brings us to a conclusion that when electrons flow in a conductor from—ve to +ve then radions and magnetons must be passed from nucleus to nucleus from +ve to—ve. The energy state of the conductor can show losses through convective and photonic (heat) losses. All energy required to do the "work" of emitting any kind of energy from the conductor is supplied by the external energy source e.g. a battery.

Maxwell is still correct in his gravity and electric field propagation resolutions. It's just that magnetic fields do not extend far enough, or electric fields for that matter), to enable the transmission of radio waves far into space, if space is a scalar. Only this new theory is not problematic in any regard and it is a "probable fit model" to describe the effects as presented. I must conclude charge and magnetic fields to be formed by virtual particles, if it cannot be accepted that ramatons are also force carrier bosons which I suggest they are.

CHAPTER 12

REFUTING RELATIVISTIC SPACE TIME CONTINUUM THEORY

The childish representation of the bending of space time by an object of supposed mass resting on a cross gridded "trampoline", with another smaller ball rolling into it to represent gravity is one of the most facile descriptions of an effect I have ever witnessed.

How can physicists (Ok you've got better things to do!) ever give accreditation to a supposition that requires an angular force (ostensibly at right angles to the space-time continuum) on every infinite plane of such? Please explain this force and where it comes from! Ah you say "The force is gravity!" Now we have the fascinating explanation that the warp causes the force that causes the warp ad nauseam. It is a circular argument and a significant dilemma.

So you may now argue that time causes the warp in space and so changing the distance/time relationship. This speculation has no empirical grounds in physics. i.e. What apart from "relativity theory" (itself being the beginning and the end of this similar type of circular reasoning) allows the constant of time to vary and enable this effect? duh! Duh! And double duh! Even being such a skeptic I may allow this to stand, albeit as 'shellacked" science, only if this new theory doesn't offer a more plausible and explanation of the nature of things.

It is far better to suppose "supernatural forces" to be at work, than to superficially twist observational facts to suit the ends, because history has shown that such forces as gravity are simply ones we don't understand yet. For example; Once upon a time there was a man who realized that the apple that fell on his head occurred by reason of a force. Up until that moment the force was taken for granted and if it was even thought about at all it would have likely been determined to be "supernatural". The fact that the fine thinker

Mr. Newton may have got it a bit back the front simply showed a deficit in full understanding. It is this lack which necessitates such mind gaming to try and make sense of a force which up until now couldn't otherwise be explained.

Fantasy and mind gaming is an essential part of theorizing but theories without substance should be tossed out within a respectable time frame!

The fascinating idea of being able to bend or warp the "space time continuum" leads to a more realistic idea that might not be so silly, but rather than bending (or even better still, crumpling up the bleeping piece of paper with a grid drawn on it) so as to somehow describe time or space travel through wormholes, my theory actually envisages the remote possibility of universal travel through black holes because faster than light speed and encapsulation and protection technology may not be out of the question someday! However it would require either stupid or intrepid explorers (perhaps both) to venture on such a mission!

For me to take on Einstein might seem to be rather impertinent. But I might be able to ameliorate this impertinence by pointing out that he has himself already refuted his own assertion re "c" being constant with his space time continuum blurb. He has also been recently proven wrong with his theory of Brownian motion.

Considering that I have no great string of letters behind my name I risk nothing in the dis-accreditation stakes and so here goes.

The problem with Einsteinian space time continuum cosmology is that of the human observer. (Who gave us the right to consider our planetary system to be the centre for observation of the universe and even for those observances to be scientifically justifiable in human terms?) Remember how recently scientists believed that the Sun and stars revolved around the Earth?* What was the cause of such incorrect assumption? Yes it was false observational "relativity" assumption. Einstein's observer should therefore be an empirical and logical observer able to see and measure without instruments, in an instant, and from outside the universe. I will address the findings of the proverbial and "magical" traveling observer shortly. *Also reflect on how Galileo was treated for his findings.

With this empirical, logical observer in position; consider him watching scientists conducting a speed of light experiment on Earth in two opposite directions at the same time with all other relevant things being equal.

The Earth is moving through space at 30 (or so) km/s. The Earth bound scientists both achieve the same result; however the "observer" sees light going forward faster than light being measured in the opposite direction by about 60 km/s. So; "speed of light" experiments on Earth only go to prove that the speed of light traveling through the universe can be different, otherwise we are have to expect that both scientists age at different rates simply because they are facing in different directions! Wow we have discovered the "elixir of youth". All we need to do is develop a computer program which tells people which way to face at any given time and they will age more slowly. Please; this is sarcasm! Have you not got it yet?

One of the problems here is that the laboratory is moving with the experiment, which is not at all being taken into account in a literal sense but only relativistically with the accompanying logical conundrum.

I can think of an objection at this point, in that one might argue that "real world" astronauts in traveling to the Moon and back should have noticed a difference in color relative to their direction and velocity in space should relativity be incorrect science. Such reasoning is not really plausible because the astronauts are only traveling at somewhere around 4km/s, which is insignificant. But it is fascinating though, that they brought back (Hasselblad) photographic pictures that appear to be quite color changed dependent on the direction of their travel! They appear to be red shifted as the Apollo spacecraft departs Earth and very obviously blue shifted as they approach. It is in fact as obvious as "the hand on the end of your arm"! Have a look for yourself.

So "Houston we have a problem" by being faced with seemingly contradictory observances, because on the one hand we have many experiments which show very little change in "c" dependent on the directional velocity of the Earth yet we observe Doppler shifts from outer space and in photographs from moving spacecraft. How can we correlate and explain the seemingly inexplicable. I will do just that shortly.

Another set of observations which scientists are trying to explain away is to do with the aperiodic time variabilities in Binary pulsars, and the strange problematic phenomena that the "red noise" distribution is not according to the Lorentzian component which only has 68% compatibility, rather it is according to a different power law. I can explain both of these.

The velocity of the x-rays is being emitted at a higher velocity from one side of the rotation than when the orbiting star is moving in the reverse direction. This velocity difference would be in the order of 0.2 "c". Also if you look at the frequency shifted gamma ray image of the Vela pulsar for instance, you will notice that the pulse is blue in one direction and orange in the other. This can also be explained away by typical means, but I suspect that it is actually orbiting a black hole; I rest my case regardless! The Lorentzian curve problem I have already addressed in a previous chapter.

I have devised an experiment to prove Einstein wrong in that he theorizes that light emits at the same velocity from the front and back of a moving object and is a universal constant, and that he supposes that light appears to remain at "c" to an observer in the object because his (the traveling observers) time dilates with respect to an outside "observer". Finally then we might get the admission that light actually travels at different velocities in space/medium, because the two suppositions are logically at odds and cannot be selectively abrogated to suit the situation. Either light is not a constant or it is.

I don't have the resources to conduct this experiment myself.

Using a speeding tracer bullet traveling at around 1000 m/s and at least a long enough circle of light fiber (A thousand kilometer round trip borrowed from your friendly "Telco" would be good!) and by utilizing optics and an appropriate oscilloscope it should be possible to measure a difference by delay of travel in the light fibre between light transmitted from the rear of the tracer bullet when compared to light transmitted from the side which can be considered in a vector manner as a stationary light source and therefore the control beam. (You are well able to measure delays of a nanosecond or so. Anyhow it's just an idea; oh and also, would there be any Doppler shift that might be measurable by spectrometer as well? To silence the critics both measurements should be achieved)

Most physicists would find this very problematical at first glance because the light would have to pass though an amplifier and this would be suspected of muddying up the results. I would like to point out that you wouldn't be measuring "c" but simply the difference in time it takes for lower velocity light to travel the distance involved compared to a control beam at "c", and I assume that amplifier and media caused delays would affect both beams similarly.

The main problem would be the losses realized in current light fibers, because of the necessity that no amplifiers can be inserted anywhere in the loop except for an end terminal amplifier, because the light path must be "optical" without "new light substitution" all the way for obvious reasons. This experiment may have to wait for extremely low loss fibers to become available.

Currently the best optical fibre has an attenuation rate of 0.5 dB/Km which (if you could live with a length of 500Kms) in that case the 250dB loss might still retain enough retractable signal to amplify at the receiving end, and obtain an empirical and repeatable result. Perhaps a differential velocity experiment would give better resolution.

Of course the old circular argument is going to raise its ugly head. Aw! Either the distance or time has changed. Get over it! You can keep arguing about what actually changes all day but that doesn't alter the fact that throughout history science has prided itself in popping the question. What is the OBSERVED change? And lo and behold! Either they calculated the correct outcome or arrived at an enigma or logical difficulty, which they then recognized would require new scientific modeling that does not abuse logic or rationally understood observances to solve the problem.

To think that mental gymnastics may lead to future technologies may very well backfire if it causes the proponents of such beliefs to be found barking up the wrong tree and the squirrel gets clean away! People have been trying to prove relativity for about a century (which is way longer than the SETI program which should have been packed up and mothballed by now also). and all the while they simply show that they are attempting to prove relativity with itself. Consequently I declare that it is improvable: Either way if logic gets throw out the window, as "sci-fientists" are apt to do: it is a stalemate. The only way out is, if another method of describing how everything works can be presented, that doesn't leave science with the hand it has been dealt. I'm calling for a Re-deal!

Scientists should use mathematics to explain enigmas not logic. For example the enigma of the following example . . . 1xo=? can be logically explained to equal either 1 or 0. However if you use the faculties of math; Then, by using the law of transposition, 1x0 can be transposed to 0x1 which is patently obvious to be 0, so by math 1x0=0 full stop! So I am aware that in certain circumstances even logic can't be trusted!

Many relativistic methods of calculation can be supplanted by classical methods. Relativistic methods may sometimes be easier and arrive at the same result, but that doesn't prove relativity in the weird sense. It simply proves the relationship of "c" within the observational parameters here on Earth, and is a mathematical model which I believe I have shown is not accurately relatable to the universe in a broader sense.

Doppler shift and light: I understand that relativity explains away how Doppler shift is supposedly able to occur without an actual change in "c". I wish to dispute this by the following reasons.

Any vibrating object moving at velocity will trace a wave (non existent) and we all know how Doppler shift works. So Doppler shift is seen to occur in this theory for this, and only this reason: Because light is traveling at a changed velocity!

To conclude that relativity causes the light to stop vibrating while it is traveling because it is time dilated to non motion, is another weird logical conundrum which seems to disprove wave theory by relativity itself, but I won't take that tack because of my belief in logical reasoning. Why is it? that science won't accept the simple reason (being more likely the right one) but has to stoop to mind games? So I conclude that if Doppler shift is observed then light speed must be changeable.

What if the "proverbial" astronaut traveling near the speed of light shone his torch onto a mirror moving at the same angular velocity? (OK then, even a stationary mirror!) According to relativity he would observe the light go towards the direction of travel at "c" because ostensibly, his time has slowed and he appears short. Now the light reflects back and he still observes it at "c" How? If his time has slowed down then the light reflecting back at "c" would appear to him to be moving at almost twice "c". Either that, or either he or the "magic" observer, possesses the strange faculty of time dilation-expansion duality whereby half of him has slowed down and the half with an eye on the reflection now has to appear to travel in reverse at about "c" to keep his observation of "c" at "c", and this as well as having himself half shrunken and half stretched!

The theory of relativity is patently nuts! What makes this crazy is that his clocks and radioactive decay rates and everything else aboard his ship are going slow as well. Whoops sorry! That's just half the ship. The half of the ship that the light is reflecting back to has clocks etc going twice as fast. Phew I'm glad that's clarified!

This generates the next question which actually needs to be answered first. What happens if you shine a torch on a mirror moving away from you at almost "c" and relative to you being the observer? If your answer is; that light will still reflect back at "c" (which of course must be the answer of a "c" constantist). Then my previous question leaves you with quite the mental conundrum? difficulty? contradiction? absurdity perhaps? Please isn't it time to give this stupid relativity theory the flick?

OK then let's expand on this: (If you are thinking that I am about to embark on the same tired old arguments that you are well aware of: Think again!)

If an astronaut is shrinking in time going forward, what happens when he looks back at light approaching from "home" and still measures that at "c"? If he is surrounded by clocks, are they all ticking at different rates? And does his "time dilation-expansion" change or does his body get shorter or longer depending on which spatial direction he happens to be glancing to perceive incident light?

Of course you may be smug in your argument that relativistic observance is different for light beams from various directions because we are analyzing the effect from single frames of reference of a "magic" observer who is able to do this.

You will then stop the argument at this convenient point, without realizing that the argument still hasn't been concluded and the scientific method is therefore being abused and the reason is as follows. The observances of all of the data including clocks placed evenly around the spaceship should be noted, and the average change in clock speed will then be calculated and concluded to be zero. Perhaps you haven't heard of such concepts as "averaging" or "interpolation", or if you have then maybe your college taught you that you may dispense with such scientific tools at will, in order to maintain wishful belief positions.

This concludes that even the "magic" observer notices no actual real contraction or time dilation. That's correct, absolutely zero. I have presented a valid reason that prevents travel at the speed of light with current technology, which has already been explained herein.

The following section will analyze the cause of the velocity of light being "c", and also solve some of the problems noticed in the universe with a non relativistic approach.

THE VELOCITY OF LIGHT AND RELATIVITY:

First and foremost I will restate that my theory concludes that the Eos is responsible for the velocity of light at every time and place in the universe by taking the universal and local area pulse by several parameters being the relevant temperature as well as GD and local GS states.

Firstly; in answer to one obvious objection which by now may have arisen in you mind.

You may have reasoned that because the whole solar system may be traveling at such incredible velocity that some sideways inertial GS effect should be noticeable by a tilt in the vertical orientation of objects such as "ourselves". It should be noted however that such affects would be infinitesimal in comparison to the masking or swamping affect of the massive GS of the Earth by comparison.

Light speed experiments have shown little velocity difference produced by differing directions of travel of the earth through space. At first these experiments were carried out to prove the suspected existence of a light propagation medium called the "aether". However the conclusive results caused scientists such as Lorentz and then Einstein to consider frames of local time reference and the dilation of time and shrinking of length in an effort to make sense of the results because they falsely concluded that light must appear to be constant to a moving observer regardless of is velocity. It must be pointed out that the fact that all experiments carried out since then have actually shown a small but existing fringe shift which does show a slight yet significant difference in the measurement of "c" with velocity of motion, but well under expectations.

Even though disproving the existence of any aether it may point to the fact that "something" must be causing reduction in the expected changes in "c" and I intend to show you now that the answer is not relativistic.

In fact how it can be concluded from noticing a lack in experimental changes in velocity that a person's time and length must have changed is beyond the pale. For crying out loud they forgot that the whole experiment was moving through the universe, and therefore in so measuring no change in "c", the scientists were ironically measuring a real but unrecognized change in "c" which would be observed in light being emitted into the universe. If "c" is being emitted from atoms at the same velocity from any source and in any direction on the Earth. This observed change in velocity is what we would expect.

To make this phenomenon "relativistic" relies on the assumption (that they also derived from this flaw in reasoning) that "c" is constant everywhere in the universe, which we know is not the case because of all the Doppler shift information which is observed right throughout the universe. In fact this is the only way we can tell that the Voyager spacecraft are slowing down. Yes; by emf Doppler shift!

By barking up this wrong tree, they could then perceive a relative time problem between a fast moving astronaut and his relationship to earth time. However if the astronaut was in a similar "gravity"* his measurement of "c" and his time and shape will remain the same but then the velocity of light emitting from his ship will be different in relative spatial directions. We have already seen that there is a very different reason which prevents high velocities of matter in space. I.e. Inertial GS, or space drag. *His atomic clocks may become slightly affected by GS differences in different parts of space. This is because the Eos controls internal atomic behavior of atoms which also results in the change in light velocity. Any change in "c" caused by GS is also expected to be controlled by fourth power law.

I sense a need to re-visit the following: GPS satellites give us a clue to the significance of this effect because they are orbiting in a reduced GS zone which is proportional in some manner to their distance from earth. It is likely that they will also have some small daily inaccuracies as well because of the increased affect caused by the changing GS of the sun.

Now according to my theory: The Eos (by providing the required cmf across various tines), causes the propagation of massless particles such as photons and ramatons.

As previously explained, the Eos determines this velocity by reading the universal and local indicators of temperature, GD, but most importantly of all for this explanation; the local GS in the vicinity of the light emission point.

At any point in space with a similar GS the Eos sets the speed of light which we measure on and near earth as "c". The velocity of light by this reasoning may be slightly dissimilar in other regions.

Because of the universal readings the Eos is able to change the velocity of light at any point in space at any time, and perhaps more importantly, for any period of time.

Now we arrive at a logical conclusion that actual measured "c" being controlled by local GS is in no way relative to the velocity of any moving laboratory existing within that GS and therefore of course any measurement of the velocity of light will be the same, and as we just determined, "relativity" and local frames of time are no longer required to be postulated because there is no longer any problem to solve. Note: the experimentally derived fact that "c" changes, in a moving medium such as flowing water is an observance that is not relativistic because it is not only supported and is explainable by this theory with special reference to the pertinent section in chapter 11. This is the section which explains the apparent change in the velocity of light through a medium. This cannot be explained by standard theories, so relativism is deemed necessary. I trust that I have convincingly shown that it is not.

By my theory there will necessarily be anomalies in the measurements of distance and density of bodies in the solar system as well as the universe, and (specifically because of the Sun's relative adjacency to Earth) solar measurements.

To reiterate: The fact that the velocity of light is a measurable constant at any point in space and even on a planet with a large GS does not require relativistic explanations. But place a light source or camera well away from the GS of the Earth and the results will be different because of a different local Eos reading, and that light from a stationary object in relation to the earth will be shown to actually travel at slightly different speeds in space. Even with our galaxy arm spiral velocities the difference in "c" may only be a maximum of 1/100 of one percent, which will of course have no noticeable effect on our verticality because of any proposed inertial GS change.* This is because of the observed luminosity to velocity of a spiral arm galaxy seems to suggest that the velocity to GS relationship is also a fourth power curve function. Which by the way; agrees with what I stated about the power curve function, back in chapter three. *In any case if the velocity of the earth through the universe is incalculable we do know that it is of insufficient velocity to cause any noticeable inertial GS effects.

Any measurement of verticality is also very problematic because, horizontality is itself determined by observed gravity induced verticality.

However this velocity through space of the whole solar system results in a slightly elliptical orbit of the planets of just a small ratio of a small fraction of a percentage point. The subject of precession has been addressed in chapter six.

The velocity induced inertial GS of a high velocity spacecraft* will of course now be observed to result in a different light velocity being emitted from the front than from the rear. It will also result in different measurements of the velocity of light from light sources from various directions. *Likely to be noticed only by Doppler shift until I reaches velocities above 100km/s.

Any attempted measurements of incoming light by utilizing the radius of the earth as a "yardstick" will be rather futile because of the infinitesimal differences in "c" being caused by universal motional affects. Even noticeable Doppler shifts are often ignored misinterpreted or explained away with distorted relativistic arguments. Like for instance a 68% match to a Lorentzian curve is good enough proof isn't it?

It is a common occurrence for scientists to simply ignore data as anomalous when it doesn't concur with their pre-conceptions. A case in point is that upon the observance of a recent supernova event, it was noticed that a neutrino burst arrived on Earth some 18 minutes before the light arrived. Rather than attempt any explanation for such a problematic experimental observance it was simply ignored for while without comment from the academy until internet interest in the dilemma required a protectionist response.

The supernova 1987A was the event described above. Since then explanations have been given for the neutrino arrival time anomaly which include the convenient ascertation that the neutrinos were emitted from the core of the star, and the light was either emitted later or it was delayed for some reason. I find such explanations rather droll.

If scientists come back at me and my gravitons and declare that the gravitons from the sun arrive on Earth about eight minutes before the light because the light was delayed by eight minutes. I would find that to be one coincidence too far. One that would not even be acceptable as a plausible explanation by Homer Simpson!

All that aside, now we can conclude that the velocity of light can be different than "c" when emitted or received by an object outside of the effects of a major "swamping" GS which is able to mask most velocity related GS differences as we have just noted.

The reason that most binary pulsars do not show any significant fringe shift is because in that case the two interacting stars are actually existing in a combined GS field and this is especially so if they are orbiting close to a black hole. It can now be easily understood how this can effectively mask much or all of their velocity induced inertial GS.

The variations in Eos affected light speed changes can be noticed by fringe shifts which are in fact observed in some binary pulsars. This is very problematic and has led some scientists to once again envisage an interstellar light propagation medium (aether) which has a variable refractive index, because even scintillation effects have been noticed as well. This is all able to be explained by Eos activity, due to the previously mentioned GD fluctuations in the universe. My theory has nothing in common with any aether drag hypothesis.

The frequency of light can remain un-related to velocity in certain circumstances such as by inertial GS effects even when the changed velocity is occurring in a GS masking field. This is because frequency is related to force/energy and the velocity is variably related to the Eos.

The luminosity of spiral galaxies is proportional to velocity according to forth power law. This is caused by increased inertial GS and resultant heating of stars within the galaxy in such power law proportionality to increased velocity.

We can also answer the question of why galaxies with spirals seem to rotate at a similar velocity regardless of the distance from the galaxy centre.

We would expect the outer arms to be orbiting at a much slower rate similar to planetary orbits. However this just proves that spiral galaxies are not "orbiting" at all. In fact the centre region of these galaxies is usually about to, or actually is in the process of spiraling into the black hole or at the very least a neutron star about to become one.

This is because the outer arms are traveling at a restrained velocity which is enough for them to overcome the GS of the central black hole and maintain an orbit. However the closer the arms are to the centre, the more the velocity constraints imposed by this space drag theory cause them to spiral into the black hole. We now know what causes the velocity restraint. Yes the previously explained space drag! So now there is no need for the postulation of any kind of dark matter which has some sort of "magical" gravitational affect on the galaxy.

This is the final nail in the coffin of relativity and it also creates problems for distance measurements and age speculations of the universe. This is also compounded by the tine switching (bending) of light around massive bodies, which casts doubt on the ability of scientists to accurately measure distance by the binocular method of utilizing bi-annual Earth orbit spatial differential trigonometry because the light would be necessarily bending as it approaches the sun with an unknown value.

Note: I have written but not included a whole chapter refuting Einstein's formula $E=mc^2$ because after consideration, I realized that even though cosmologically incorrect, in our local area of space-time it works fairly well for most calculations it is normally applied to. Although not exactly correct it is a close enough approximation at sub atomic levels, because the delays in "energy" transfer in that world are insignificant, so the formula still stands. (Believe me to arrive at the formula, Einstein married gross assumptions with a mish-mash of formulae whereby (as luck would have it) he chose the right experiment that approximated actual point time reality and so he arrived at the somewhat true formula: Happy days!

CHAPTER 13

THE EOS AND THE GOD PARTICLE

The term Eos is actually derived from "Executive/forceos" which is a bit of a mouthful, so Eos it is.

This explanation is going to complicate matters more than necessary for the presentation of the theory as a whole, but I feel it is definite need to clarify this to avoid the obvious logical dilemmas which will otherwise be encountered.

Several definitions must be clarified first.

1/ Weak nuclear force: This is the force contained within a single nucleon (proton or neutron) which holds the nucleon together. It is the primary function of quarks.

2/ Nucleon or atomic binding force, or strong nuclear force: This is the force which holds nucleons together in an atomic nucleus against the proton Coulombic repulsion. This is the role of quarks also, but in this case by utilization of bosons called gluons.

3/ Extreme near-field force: This is any force which only acts within a nucleon and does not transfer anything beyond its borders, whereas near-field force finishes at or near the boundary of the outermost electron shell. Strong nuclear force is an extreme near-field force, while binding force is a near-field force. Quark and some bosonic forces are extreme near field force

4/ Sub-quantum: Force quantum numbers measured in integer steps of three sub-particles. Sub-sub-particles are not quantasized. Quantum integer numbers also reference with, electrons and graviton packets such a photons and ramatons.

5/ Bosons: Sub-particles of quarks and greater particles: There may also be three smaller sub-sub-particles which can be enjoined to make up fundamental bosons. These are actually particles of pure force. The characteristics of the particles translate to static forces which are charge, magnetism and bondage (binding force) or not. These may be contributive to bosons of greater size and which are—charge boson or particle; magneton boson or particle and the atomic binding force boson or particle all being combined to constitute the graviton particle. These are neutrino, magneton and gluon respectively.

Be that as it may; the known bosons maintain force parity when confined in a nucleon or a graviton. They do not attract or repel each other, they form equilibrium. They only attract or repel bosons in other nucleons by tri-active near-field force effects. I have in my possession, diagrams that show boson combinations that allow "non-magnet and magnetized matter objects", boson arrangements which still maintain the strong atomic force in both cases.

Bosons especially gluons and even electron orbitals all combine to hold atoms together. It is likely that strong magnetization will change physical characteristics of a material to a certain extent. It is theorized that static magnetism in materials can only be maintained within the window of Eos operation parameters.

This same model shows how a magnetic line of force can cause poles to rotate and shift spatially within nucleons, while still maintaining nuclear bond. This was a necessary condition for the explanation of electrical generation by magnetism in chapter seven.

The three sub-sub-particles combining constitute the first particle of matter not by magic but by pertinent translation to the dimensions in which we observe matter to exist. Matter is therefore not the existence of stuff or not. It is caused in all of its manifestations by the inter-working of forces in all of the dimensions as the case may be. Note: It is of extreme importance for the gravitational theory herein that it can be concluded that the Eos maintains the "matter" content of all objects with nucleonic constructs !

Bosons must not be referenced or equated to have characteristic of matter that we observe on the larger scale. They are not heavy, hard or soft, brittle or plastic, hot or cold. The only characteristics of matter that they possess are that they occupy space; they can move and exert force.

Magnetic poles and charge poles (signs) don't actually exist. We name them and give them their attributes only to make sense of the observations that we see. Sub-particles move and operate in and dimensionally parallel to the Eos in the manner determined by its laws. At temperatures above and below the "operating range" of the Eos strange things occur, many of which are explained within the pages of this book. By their respective actions and with the introduction of velocity/vibration, these bosons contain an attribute called "divergence" force. I will now explain.

The Eos oversees all of the dimensions listed in chapter one from dimension five to twelve. It transmits force sub particles to wherever they are required for their application, to provide the necessary equilibrium balance of force. The Eos measures both the overall temperature and gravity pulse of the universe and of individual atoms. It is responsible for keeping objects separated or bound. This occurs instantaneously because the sub-particles have no mass. I will give an example.

When a photon is emitted by an atom to a tine, the Eos immediately transfers force particles and velocity "c" to it which as we shall later come to understand is related to a particular energy state requirement of the universe through divergence. Extra force particles do NOT add to the mass or density of any object. They are given and taken at will by the Eos which is not subject to any universal law because it is causative of those laws. The photon is also held together on its tine by the Eos.

When two photons collide, (pass through each other) they do so by shifting dimensions*. When this occurs, energy is transferred from the photon to form a graviton/s. The Eos confers a velocity "y" to the graviton which is also subject to divergence. The graviton in the gravitos dimension now has energy and divergence force. *During normal non-plasma photonic "collisions" the photons maintain contact with the brane during this process and they then continue along the same tine.

Force and energy: Two of the most unusual forces which almost seem to create energy from nothing, are friction and impulse. They are actually two forms of the same thing.

A graviton has the force/ energy content of the smallest photon. A photon (quanta stream packet of "graviton sub-quantum sum" force particles) gives up force particles and so loses vibrational kinetic energy to form a graviton/s which is then converted to vibrational divergence force in the graviton. The motional change related "energy released" by the force of graviton transition "friction" may eventually be re-converted back to photons in another nucleon or photon stream. In the low intermediate state the gravitons are reduced to the static force sub-particle components until sufficiently high energy levels (meaning addition of sufficient extra bosons) allow recombination of the sub-particles back to gravitons. A graviton can therefore become a dimensionally shifted sub-quantum of tri-potential weak force or a basic quark.

We have finally declared the mechanics though which graviton transitions cause the effects of gravity and mass by transitioning through a specific graviton construct within nucleons.

A photon being received into a nucleus is immediately disposed of in any manner and as soon as possible, either by re-emission, BBR or convection. The latter being a sharing of gravitons or sub-particles with its neighbors. A rise in temperature will also signal the Eos to act in order to keep force/ matter balance requirements in check. Because of this and other reasons so stated, the Eos cannot act as instantaneously as it would like, and temperature differences in matter throughout the universe, is the result. This is also why gravity anomalies can be observed in the universe. These variations are often called gravity waves. They are not the cause of gravity but are simply a barely observable effect of such.

The velocity or vibrational frequency/amplitude of massless particles has nothing to do with momentum and they can be postulated to possess wave particle duality but only in a mathematical sense. I do not see any need for wave function to explain the operations of these particles because of the multidimensionality and subsequent ability for them to occupy the same space at the same time. In such a case a graviton would be the same physical size as the particles that it consists of, which is why we can still consider it to be a boson. In this case uncertainty principle is invalid because we know exactly where the particles are regardless of their movements if we know were the graviton is.

The vibrational velocity moment is a time delayed indictor of the energy requirements of the universe with local variance. In the case of gravitons it is an indicator of the "frictional" force reserve capacity contained within them. The reserve capacity of the transferable energy content of a photon and other energy stream packets, is in their vibrational amplitude and frequency as well. A graviton however can be seen to possess a negative divergence velocity force duality, whereas a photon has a bi-divergence vibrational force duality. This just means that a graviton can lose velocity during transitions whereas a photon can only lose vibrational amplitude, regardless of particle content value.

Force in all the bosons so described as strong force (gluon), magnetons and neutrino bosons, is by bi-divergence duality with velocity. I. E. When they are spatially motionless with relevance to each other they can exert no force.* All of these bosons therefore (including photons) appear to have a "form" of momentum even though they are massless. This is simply a function of the Eos.

I will explain later how the Eos confers velocities and force sub-particles to these greater particles on a more or less instantaneous and continuing basis in response to the "pulse" of the universe and local and motionally generated inertial GS as well as overall GD "readings". I will however be no longer be including any particles below the level of gravitons, photons and ramatons in the following address. It is sufficient to understand where force comes from, because it cannot be left to the assumption that it comes from nowhere! *In the real world all force fields are vibrational when not propagating and it must be remembered that nothing can occur without the operation of a force.

Now this can leave us with the conclusion that the Eos appears to possess intelligence. On the contrary: The Eos is simply the cosmic dimension of order in stillness. It has been awakened by the creation of the universe and all of its seemingly calculated behavior is simply the Eos vainly continuing its attempt to obey the first law of the cosmos. i.e. "Repair me; this instant!"

It is trying to do so, but real time is complicating matters by stretching the instant. Of course the cosmos has no inkling of time so it just does what it does. In fact this makes it really appear to be quite stupid, because it is seen to be working against itself by attempting to pull the cosmos back together in two conflicting and self defeating ways.

First of all let's analyze what the Eos actually does. First and foremost the Eos is responsible for the operation of the fundamental and basic forces and velocities of particles.

1/ Binding forces; being both strong and weak nuclear forces: These forces which hold atoms together, are the outworking in the universe of the overall cosmic force which is in instantaneous yet continuing operation. This is in the attempt to pull the cosmic rent back together by cosmic law. The sub-particles of these forces i.e. gluons, magnetons and neutrinos have "strength" variability, but will not move spatially or dimensionally without the required sub-quantum number of value. At ground state near zero k they also will not move until another similar although sufficiently strong particle is able to take their place to ensure the structural and matter value integrity of the larger quantum particle they reside in and are hence affecting. In a different sense; any quantum particle may contain force bosons greater than one quantum integer step, but only an integer step value sub-particle and above, may be released from the atomic construct in any case. i.e. a photon. If a photon "quanta matter" value leaves the nucleon it will be replaced by the Eos either instantaneously, or (if in the bowels of a media) by nucleon transfer at "c".

Having stated all of this it becomes conclusive that the gluon is responsible for the strong binding force within nucleons but with conditions which have previously been explained, and the weak forces are caused by the sub-particle bosons themselves. A quark will always maintain at least one graviton value of force/matter. The containment by quark of more than one sub-quantum is noted in some theories as color charge differences. Because of BBR the graviton content of all atomic objects at STP remains constant on average over time, and regardless of temperature within the effective temperature range of the Eos. Because only one graviton within a nucleus can occupy he gravitos dimension at anyone time, this will cause both very hot and very cold AMOs to exhibit the same gravity, mass and motional effects.

2/ Magnetic force. This force is often taken for granted as requiring no explanation. However I postulate that it is one of the basic bonding and motion causing forces of the cosmos. People often wonder about what causes a magnet to be. My trite and simplistic answer is . . . The same thing that causes an object of any description to be. It came from the cosmos.

The particles of magnetic force are magneton bosons with neutrino sub-bosons.

3/ Electrostatic charge force. Ditto . . . sub-particles are neutrinos.

4/ Separation force. Now this is a force which is not often recognized. This is a force that keeps universal matter apart until it tends toward a high or low temperature in which case it is deemed to be moving from the state of equilibrium towards the cosmic state as the case may be, and the Eos loses (relinquishes?) control of it.

5/ The cmf and gmf. These are cosmic forces, and are determined by the Eos in response to the overall "state of the union". These provide velocity to gravitons and stream packets.

6/ Gravitational force. This force is as explained in preceding chapters and the velocity of propagation of the force is again controlled by the Eos, and it is subject to cosmo/universal laws.

While sufficient divergent energy remains in the universe, the Eos will continue to exhibit seemingly self defeating behavior. On the one hand it utilizes divergence force to cause BBR and other energy emission, and on the other hand it is attempting to give heat energy to atoms via the gravitos, and because energy enters by gravitons and leaves by particle steams at the current rate; we have equilibrium. (The word "energy" may be used for description where specific force actions are not a necessary criteria for the explanation of motional change, but energy otherwise means heat).

The Eos however, only has a narrow window of operation because of a couple of reasons. 1/ loss of power (capability). 2/ loss of independence (authority).

1/ The loss of power has occurred because of the introduction or creation of time and distance and universal matter objects.

2/ The loss of independence is caused by it's effects being able to be overcome by local forces. Some examples are . . . Binding energy can be overcome by temperature gain or loss. E.g. burning. Magnetism can be overcome by the same. Separation force can be overcome by temperature. E.g. melting and combining. Electrical charge and other forces can be overcome by local far-field charges and emf's. E.g. drop a spanner on terminals of a car battery and note what happens!* The Gmf and emf are able to be overcome by changes in universal divergent energy. *Don't do that at home. I'm a professional!

So now we can understand that the Eos (though playing a crucial role in the operation of the whole universe) is not intelligent and it is also not all-powerful. In fact the Eos is subject to the first pre-fundamental dimensions. Yes; and surprisingly the unsung heroes of the saga are the first four dimensions which are commonly called space-time. The other listed dimensions are also subject to them.

In the cosmos, all the other dimensions had no idea that space-time even existed, because the cosmos was not required to "do" anything. The other fundamental dimensions and forces acted statically and instantaneously to keep the rest state as previously described.

Now herein lays a conundrum or two which must be addressed.

1/ If the Eos acts instantaneously with force, then shouldn't all of its actions throughout history be deemed to be occurring at the same instant as well?

2/ Shouldn't the cosmos then be declared to be of infinite size.

The answer to both questions is; that the questions and answers are all relative and there is no possible answer that will satisfy the conclusion of all possible observers. However we should perhaps take notice of the fact that the Eos is performing an almost infinite number of time delineated yet instantaneous actions of varying duration throughout the "brief history of time". I couldn't help myself! To explain further: The actions of the Eos may have instantaneous rise and fall times yet a non instantaneous duration because of time and distance constraints within matter.

At night when all the world's asleep the questions run too deep for such a simple man . . . *Supertramp*

Is there anybody out there? . . . *Roger Waters . . . Pink Floyd*

CHAPTER 14

FORCE ENERGY AND MATTER

Force at the fundamental level is great or small but the problem with classical physics and quantum physics is in the perplexing enigma caused by the very large seeming to have the very small force and the very small having the greatest force of all. These forces are gravity and strong nuclear force respectively.

The other problem is that an atom also has other very weak forces which seem to be able to overcome extremely massive external forces.

The attempt to address these problems has resulted in all manner of theories. In particular various string theories, one of which has come tantalizingly close to this one. That theory is M theory.

The problem with all of these theories is that they are all attempting to correlate with relativity and the modifying and or inversion of time in attempts to bring their proposed dimensions back into the three plus one dimensional fold.

With my theory we could call my non-existent lines etc strings, but it still would not be a string theory because it is non relativistic.

The complete theory explains the cause of gravity and why it appears to be weak, because it is not caused by the same thing that causes the other forces.

Strong and weak binding forces are as there namesakes suggest, caused by force sub-particle density in the required regions. Atomic bonds by valency are caused by the vibrational harmonic arrangements of the force particles to form patterns as Hilbert space sets interlocking with corresponding sets of other atoms.

Force particles are controlled by the Eos and the other dimensions so involved and they have attraction or repulsion between themselves as previously explained, but in outcome driven ways.

This fact along with antimatter event horizons being on branes is described in another theory of quantum physics as color and strangeness respectively. The force sub-particles are much smaller and below mesons and kaons which are actually expected to be complex particles consisting of gravitons (which have combinations of three force particles as described)* which then form the quarks of various strangeness and color, which in combination can then cause "color variation charges" and "spin" (rather eigenspace orientation). *neutrino, gluon and magneton force sub-particles.

Pauli exclusion principle has no relevance at these sub levels yet the sub-particles will still exhibit sub-quanta characteristics in that they have objective force definition which can be resolved as energy definition by relatable actions such as BBR and so they still have actual existence value. This is what enables them to transfer these force "quanta" to quarks. The force quanta are the actual tri-combination of the sub-particles themselves called gravitons or (where captive in a nucleon) a fundamental level quark.

Such particles as gravitons, quarks, mesons, kaons and hyperons in particular exhibit this multi level symmetry and anti-symmetry which is directly translated from the original cosmic symmetry theorized elsewhere in this book.

Supposing that every particle is available "on site" in real time and at creation level temperatures and gravity: As matter becomes built like a Lego structure from these sub building blocks the Eos causes them to create more and more complex structures and begins to cause the force sub-particles to utilize the branes and other dimensions to cause motion and energy release beyond strangeness and color to combine forces (no pun intended) to create more specifically energetic and structural particles which in the end results in matter which we can observe. This creation of matter is impossible in the currently existent "normal" universe because only the Eos is able to act instantaneously but the dimensions responsible for interaction are restricted from performing such creation of matter by space time, but prior to that when al existed as the cosmic black hole the whole cosmic system was capable of interdimensional eigenvector quark transformations. This could only have occurred in the universe before graviton motion and ensuing "impacts" occurred and it has resulted in the periodic table of elements.

Graviton absorption into a nucleon results in a slight quark color transformation. Graviton transitions results in slight boson loss in the graviton, causing a slight shifting of the eigenvectors of the charge and magnetic dipoles which translates to a nodal shift in the electron orbitals. This in turn results in an energy swap to the next atom in the direction of graviton travel and so causing heat convection. Also the transition translates as a force against the next atom as well because of the greater shift in the weak binding force eigenvector in relation to the minor shift in the strong force vector, thus causing a change in the Hilbert space set of the whole atom which in turn affects each nucleon. This is all dimensionally related as explained herein.

This results in two things. 1/ Gravitational and motional force by drag effect on the nucleonic quark/gravitons, and also the creation of a magnetic disturbance in the atom. With the enormous magnitude of graviton transitions in most cosmo-universal bodies a magnetic field is the result. This of course is entirely dependant on the dimensional status of the elementary makeup of the body according to the God code.* A graviton only loses velocity and not bosons because the Eos instantaneously re-supplies it. But it does lose velocity because of graviton-quark interaction as explained.

It is probably because bosons in the nucleon and in the graviton have an attractive force which causes the velocity loss. The nucleon gains vibrational velocity, and the transiting graviton loses linear velocity and directional eigenvector change, and the parity and energy conservation offset is zero.

Whether this is the reason or not, there is a negative inelastic collision involved and the loss of motion of the graviton is transferred to the eigenvector motional change in the nucleon dipoles as stated. Transference of spatial graviton velocity to nucleon vibrational and spatial velocity by reason of forced containment of sub quantum energy is necessarily the case.

Other matter/force restrictions may prevent this nucleon eigenvector change from occurring and the force is vector shared with other atoms in the object being transitioned, but because of the eigenvector shift of the quark being transited, the resulting directional effect is therefore in the vector relatable direction of the transiting graviton.

Of course the effect on an individual nucleon or atom would be immeasurably small and is herein postulated to be less than sub-quantum value or even atomically speaking; infinitesimal. However one should ask the question. Would an atom in a vacuum still fall by gravity? I predict that it will; whereas a boson will not.

This theory of graviton induced gravity predicts that if a perfect experiment can be performed in a perfect vacuum here on Earth, whereby you take two objects of widely dissimilar density (specific gravity) but of similar volumetric proportions; that the object with greater density will actually fall at a greater rate of acceleration than the other. This difference will be very small but it should be measurable and also relative to specific heat in a possible inverse manner. (There he goes again arrogantly refuting Newton! OK questioning anything and everything has scientific relevance: So then why don't you prove me wrong!) The trick is to ensure an absolute vacuum and perfectly simultaneous release.

All the Eos can do is what it does best, keep matter energized and moving towards the BST.

The dimensions interact via the branes and are subject to directions from the Eos which receives both external data from cosmic law as well as from analysis of sub atomic formations. The dimensions then direct the particles, acting from the bottom up to engage in the behavior which is programmed into them by cosmic law.

The original structures which existed are cosmic structures and they cannot be replicated by creation in the universe. The universal nucleon structures must be recognized as already existing. The Eos can only rebuild structure dependant on certain criteria, being 1/ temperature 2/ The availability of necessary quantum particles. Without these two criteria being met this theory suggests that decaying particles such as kaons may actually decay to an observed nothing only because they have nothing to attach to. They have not actually decayed and energy conservation has not been dealt a blow, because they have been in whole or in part been instantaneously scavenged by the Eos and their energy will be reconverted back to form more force sub-particles and the force-energy matter cycle will continue while the temperature remains within the operational window of the Eos.

If that window should close the universe would rapidly cool to near absolute zero because the speed of light and field waves would soon become zero and no energy would be transferable by photons and most of the gravitons in the universe would soon be captured by matter and remain as rest mass potential energy until the GD was so low that the universe would devolve into a mist of very cold atoms which would finally be unable to emit or convect any energy whatsoever!

Symmetry and anti-symmetry are not self deleting they are conditions of matter which could, if left to their own devises have resulted in an anti-universe. The fact that they have not is due to the Eos defaulting in one direction which cannot be reversed because of energy loss over time.

This is not a scientific paper and because it would take a team of top physicists decades to comprehensively formulate this theory, I can only present a simplistic overview of the construction of a nucleon and the dimensional shape shifting enabled by brane crossing to various dimensions, which may be interwoven within the nucleon dependant on its elemental nature.

To engage you in some reiteration; please remember that a graviton consists of a number of multiples of at least three individual force sub-particles, namely bosons: These are z-bosons, neutrinos and gluons. Quarks and leptons in turn consist of gravitons which by consequence are able to have quanta step variable force and also to be colored and given strangeness by determination of the Eos in conjunction with other dimensions as stated, and also being very dependant on temperature. Other particles such as hadrons, mesons, kaons and hyperons at higher matter levels consist of quark, gravitons, neutrinos and other force bosons.

All fermions consist of these sub-particles also. Bosons may or may not have anti-particles dependent on their dimensional relationship with the Eos which as I have stated was capable of creating either a universe or "antiverse", but once it toggled in one direction, the condition became irreversible and we have what we have; which is a matter biased biracial universe. The Eos however can create and absorb antimatter states as transitional states to facilitate matter crossing of branes.

Two or three major deviations from standard particle theory are as follows: A photon is considered to not be a fundamental boson even if it is the same "size" and as a boson. Only the force sub-particles and gravitons (by being the fundamental carriers of these) are bosons. A photon is a carrier of gravitons and may perhaps be called a P-BOSON. The other deviations or additions are the postulation that a magneton is a force boson interacting with strong nuclear force gluons, and a radion is an R-BOSON similar to a photon. These P and R particles are actually dimensionally shifted boson streams and yet (by multi-dimensionalism) they may be only the same size as the particles they consist of. I will refer to these as PR-bosons.

Sub-particles are not limited to gravitons. They are utilized in every other manner which this theory postulates. Leptons do not consist of just gravitons, they consist of standard gravitons and individual and other combinations of bosons. A fundamental quark/graviton is also a boson by this theory. Higher level quarks are not because they are in a intermediate state between sub-quantum and quantum levels. Within protons, additional bosons may exist in other dimensions other than the gravitos, and this is fundamentally contributive to the natural characteristics of the elementary atom.

The decay life of particles, bosons and sub-bosons is temperature related such that in extremes of temperature such as that which exists in stars etc and the very cold regions of space, such decay times can be tersely different than those which are observed at, not only STP but at the pitifully constrained temperatures that we humans are able to observe.

No problem exists by single dimensional measurement of the size of an object with regard to the forces exhibited. Within objects of different sizes but of the same element, the forces involved will be directly proportional to the atomic density. This is because the objects have no intrinsic mass which would be deemed to imply that a far more massive AMO would require less binding force to hold it together because it would have a much larger gravity acting upon the atoms in a "crushing" manner in proportion to the "depth" within the body,

Whether this could actually imply the opposite I really don't find necessary to address, because my reasons that objects have no actual "mass" are already clear.

Any other discussion of force comes under the subject of the mechanics of force which is well explained by classical physics and to a different degree in other parts of this book.

The forces are not intelligent. They have done what they have done, and do what they do, and that is why existence is as it is and that is all!

The interactions of the universal forces are by cause and affect which because they have arisen out of symmetry; they cause limited patterns and symmetry within chaos. This is simply because of the laws that the dimensions are controlled by are interactively based on a cosmo-legal symmetry.

It has been previously stated that nothing occurs without a force being applied. This then leads to a subsequent chain of events. These event progressions are often treated as linear and open ended. This approach leads to the "unreasonable" conclusion that the beginning and end of the occurrences can be thought to extend both forward and backward in time ad infinitum. This in no way supports relativity, rather circular eventuality.

This then leads us to take a logical approach and reason that the events must occur in a closed and circular fashion, such that for example event "a" causes event "b" which causes event "c which in turn causes "a" to reoccur. This simplistic approach is not addressing any value changes in terms.

The forces and events such actions cause may be many and convoluted to the point of seeming chaos. The law derived from this is: The number of possible events cannot be infinite or the chain becomes linear and not possible.

For the requirement of non infinitum to be met, then the steps in the event chains must be incremental with time and value. This "law" is a cause then of the theorized quantasized steps in all processes involving force above the level of the force particles themselves, whether currently measurable or not.

Within atoms, the quanti are divided by integer steps then, even at the fundamental levels. Enigmatically on the large scale they may appear to exhibit gradual change. This is because the small finite steps are lost in the size disparity and are mostly interpolated in observation.

This is why I envision the formation of Hilbert space containment cells for various packages of force particles based loosely on quantum harmonic oscillation theory. The smallest quanta cell is a graviton. When gravitons combine and reach a certain integer number they may jump to the next and larger cell to help form a photon or ramaton. These can then form larger cells which make up stream packets as well as nucleon quantum number value sets. Non integer values are infinitely possible within all the particles, but they are unable to be processed or emitted except by BBR, yet they are able to be internally utilized by the particles they exist in.

To underscore this: Many sub-particles remain to make up nucleons but they cannot be emitted to other nucleons unless they are in complete packets, but this is legal at only around STP.

There is one cause/affect paradigm which appears to be open ended, but which by the law must turn out to be circulatory. This may seem to be of no consequence to not know the "cause and end of everything"! But the law declares that the hyperverse has been made of stuff which must change from one form to others in a continuing cycle.

The laws of the hyperverse (cosmo-universe in our case):

1/ Matter can neither be created nor destroyed. It may be transported with varying velocity and density and at any determined time may be converted from one form to another.

2/ Cause and affect must be circular with a return over time which may be indeterminate, and it must occur with less than an infinite number of theoretically measurable steps.

3/ Uncertainty exists in regard to the rest state or not of any object including a force particle.

4/ Notwithstanding "law three": A force particle at rest contains potential force. Conversely a force particle in motion contains kinetic force. A force particle may exhibit force even at rest. Force plus motion results in energy.

The smallest form of matter is a force sub-particle or sub-boson. These sub-bosons can be transformed into other sub-bosons depending on the dimensional and temperature status of an atom or space. Bosons with pole or charge sign do NOT attract or repel each other inside a nucleon, or other PR boson or graviton. This is because charge sign and magnetic dipole sign does not actually exist. It is simply what we observe when bosons attempt to align to their predetermined force lines. (Strings I guess?).

Energy is a virtual object in that there is not actually any energy "mass" equivalence; neither is there any real energy "matter" equivalence except for the action of force which consists of particles of matter. Motional energy is simply the observablity of force to motion.

The term "quantum energy states" must be treated with a modicum of suspicion then, because nothing can provide such defined "packages" of value except particles with defined size and value. Virtual value energy particles don't exist, so "actual" particles of matter must be the reasonably evaluated substance. We should remember that a particle of matter has no "mass" but it is proven to be an actual object of matter because combinations of such particles develop matter of observable size and characteristics. Significantly no energy particle has ever been postulated to do so with any plausible mechanics. Any supposed energy matter transmutation must by reason occur through some unspoken mechanical magic!

The contrary position here is that everything actually somehow does consist of energy "stuff" and then we end up with a very short and circular argument which supposedly has scientific underpinnings! I have shown such a belief to be based on a questionable conclusion. In fact you will notice that because of this I have made a simple objective change to one of the laws of thermodynamics. This is stated in law "one" above.

This is because of my proposal that "force/matter" is the cause of energy being the result of motion of the particles, but only thermal energy and not mass-energy in either potential or kinetic form. Ask yourself: what is virtual force and virtual energy? Physicists including Einstein have had, and still do to this day, great difficulty in considering the possibility of the existence of such. This is why modern physics is searching for sub fundamental particles with great vigor. I would like to respectfully suggest that they back up a little and bark up a quite similar looking but fundamentally different species of tree.

Potential energy then is actually virtual energy and contrary to existing as such, it is simply the potential within force particles to be able to be moved or undergo motional change by the interaction of other force particles or their larger physical constructs such as an arm pulling back the rubbers on a slingshot. They may also have the potential to be emitted as radiation.

When two opposing and equal force particles act unidirectionally against each other in theory, twice the force is being exerted but no energy is observed because there is no motion. When motion actually occurs the energy recognized is the real (not relativistic) value of the motion of the particle times its matter value which is mistakenly called "mass". For particles with no mass then we can arrive at the simple formula $E=v$ (velocity) which for small particles we can equate (as has been already explained) to be close to $E=mc^2$. What this is really saying is that if a particle is determined to have a velocity of c^2 then its mass is "m". So "m" is actually potential velocity and nothing else! Hereby we can truly prove that "m" has nothing at all to do with intrinsic mass, weight or gravity.

Any declaration that energy is able to be converted to matter or mass is erroneous. I assert an affirmative No! (double positive or not) Energy is the observation of the force caused, time delineated step/s taken by bosons and larger objects to be formed into other particles consisting of those same bosons in differing arrangements such as photons etc. Our problem as humans is an age old one; that of drawing false deductions from observations. This is similar to the conclusion of objects being thought to contain "mass" because we can "feel and see" what we think is mass and energy, so that is what we call it. E.g. weight, light and heat are the obvious candidates for this.

Kinetic energy is also a misnomer; it is actually force particles in motion. This also infers that "force particle transfer" requires velocity caused by other bosons. The vibration of higher level particles and quanta is caused by the time delayed elastic process involved with force particles acting against each other in keeping their distance and relative position with regard to other bosons and or their interaction with gravitons and other bosons with linear velocity.

The internally interactive sub-fundamental sub-particle cosmic force is a derivative of the cosmos called cmf which is attempting to cause all bosons to move away from the universal state and back into the cosmic state. This subject and the constraints on such impossibility have already been addressed.

Having explained all of this at some length; I must now state that for most intents and purposes, the standard descriptions of energy and the formulae so used, remain acceptable. I am simply (I hope) describing a different mechanics which is attempting to tie the operation of the three objects of force, energy and matter together, and without the usual difficulties of unreasonable logic, and this is a theoretically paradigm shifted approach.

A real world example of the processes involved may be described by say a baseball hitter striking a ball which flies off somewhere into the atmosphere. To enable the ball to be accelerated in such a manner there is a necessary release of "energy" in old school terminology.

This is actually caused by the motion of force particles acting against each other within and via their various constructs (being by extension) in his body the bat and the ball, even extending to the ground beneath his feet and the surrounding air and space. This results in the subsequent motion, actions and reactions.

Because this occurs in real time there are consequent and varying time restraints placed on the motion of the force particles within all of the physical actions in the example. It takes little time for the force particle transference of velocity at the point if impulse (impact). This results in the rapid change of motion of a multitude of force particles from the bat to the baseball and it creates a hot spot on the bat and ball as well as heat convection and emission from the batter, and also into the ground etc, and it takes further time (because of extreme overall elasticity in the action and other reasons to be forthcoming) for the heat energy to stabilize in the environment. Note: I have avoided words such as acceleration etc in the interest if simplicity.

It has been consensually and matter of factly assumed, that the hitter has used up energy in the action. In actual fact he has really lost matter which must be replaced by matter. i.e. food! This matter loss is the amount of matter lost to the environment during and after the actions by force particle "matter" emission, convection and transfer.

So then it is the bosons or force particles which are matter and not energy, which is simply the measure of the velocity of the quantum packets of "force" matter doing work. Now to a different subject: Strong nuclear force.

All I am going to say about the strong nuclear force is: If you are in any way a forensic physicist and not a university clone; Then why not compare the inverse cube law of magnetic attraction against the inverse square law of Coulombic charge repulsion and get an idea of how once a nucleus is formed that other nuclei cannot get close enough for the magnetic force to overcome it. Remember this is all being addressed at STP.

The shattering of nucleons to create fission is only possible in the larger atomic (mass, weight, matter, or whatever) atoms because the strong nuclear force is weakened in an overall sense by reduction of magnetic inverse cube law in relation to charge repulsion, which is further dis-enhanced by the multiple electron orbitals of such atoms which renders them as neutral objects to other atoms, The weakened strong nuclear force causes them to be rather unstable.

I can envisage a problem with physicists blithely combining electromagnetism as one "unit force" and calling it one of the four fundamental forces of interaction.

When dealing with an atom the electric charge force is often acting with repulsion at the same time that magnetic force is attracting, so how and why should they then be declared to be one force? I contend therefore that there are five recognizable forces.

I also could present the case that strong nuclear binding force is not some sort of residual force left over from the strong interactive force that binds the parts together inside a nucleon. That interactive force is the basic function of gluons.

However it is a fact that magnetic attraction can exhibit by cubic and possibly even fourth power law over distance, and that the Coulombic repulsion law is by an inverse square law. This leads to something which may have been hidden in plain sight.

Disregarding differences relating to size, there are three forces either within gravitons or acting singularly on nucleons in a nucleus. The magnetic force is very strong in attraction to adjacent nucleons but weakens very rapidly over distance*. The Columbic repulsive force (caused by protonic charge similarity) is too weak to overcome the strong magnetic force but at a certain distance from the nucleus it becomes the dominant force. *For this magnetic force to have a case for weak binding force contention, it may be that the graviton has flat sides in some geometric arrangement or another, I cannot positively conclude any particular shape but I can imagine a square as being the strongest contender. There is a third force acting, which helps in a small way to keep the nucleus together i.e. gravity. It must be also taken into consideration that neutrons also have a magnetic dipole and they rely on their proton for Coulombic charge repulsion. This then provides magnification of the force.

If we consider ionized atoms without electrons (which we know have some minor affect as well) we can analyze this probable nuclear behavior with the following example.

Imagine two magnets stuck together by mutual attraction with two opposing rubber bands representing Coulombic repulsion of all those positively charged protons. The magnets win the tug-of—war "hands down". If however we lever or force the magnet apart*, a point will be reached (which if you think about it we could call a toggle point because it works both ways). where the rubber bands have a strong enough force to overcome the inverse cube decaying magnetic force, and in an instant the magnets fly apart and are unable to be brought back together except by a greater external force that can overcome the Coulombic repulsive force, and at this moment in time gravity is too weak to do the job. *This is how nuclei are "split" during fission; notwithstanding that gluons have some definite mediation in the strong force interaction.

The previous explanation was hypothetical but I will propose a different and more plausible mechanism shortly.

At the first instants of creation gravity became a very extreme force for long enough to thrust nucleons together and become bonded by gluons, and then it decayed. If this is true; how can we then explain large masses of similar atoms?

At the creation event, Hilbert space sets were being caused by electromagnetic force interference, (Yes the two forces of charge and magnetism can and do interact) creating nucleon positioning patterns which would have had to have been acting in aperiodic symmetry* and so nucleons were forced together in similar sized groups. Everything else followed as previously described. To put it simply: At the moment of creation, the universe experienced incredible temperatures of varying magnitude causing random diffusion of cosmic matter, and as previously explained, nucleons developed from their cosmic counterparts. *Because of the theorized pre-existing cosmic symmetry.

To simplify and readdress this: The forces holding nucleons themselves together is by the arrangement and quantity values of force bosons!

I have beaten my brains enough to get to this point. What I would like to see is some real physicists running with this, and hopefully enabling currently unheard of technologies to be developed.

This kind of paradigm shift may seem pompous or frivolous and unnecessary. My answer is; that if any technology of for instance "transportation of matter, or antigravity hyper speed drives" can ever be contemplated, the science would need to be exactly understood in the first place. Notwithstanding this I realize that many technological discoveries have come about by fluke and experimentation, but true scientific advance may be undervalued at our own peril.

Science is not really interested in why things are, rather how! The reason for knowing how is to perhaps enable positive interference in the processes even further than we have historically been able to. The reason for the quest for a formula for everything is to enable exact calculations of probable outcomes of definable human interference in such processes.

The only way to arrive at the formula is to correctly analyze the behavior of each dimension with its interaction with other dimensions. This may be even more problematic than trying to derive a formula which will result in a computer program that can enable totally predictable weather forecasts and weather system tracking accuracy.

Regardless of how daunting the task may have appeared, the inherently obvious problems have been no impediment for intrepid meteorologists. These leading scientists have persisted in palpable frustration. But regardless of the failure of many and various models, they have in true scientific fashion amalgamated substantive algorithms, and even though not having yet arrived at the "holy grail" they nevertheless have models that for the most part are becoming more reliable as time passes.

So I say "Take a look at my girlfriend. She's the only one I' got!" . . . *Supertramp*.

CHAPTER 15

FROM COSMIC RAYS TO THE FORMATION OF THE EARTH AND OTHER STUFF

Cosmic rays are higher energy photons (perhaps the highest) that are probably formed by "gamma-gamma photon" or other collisions. The cosmic photon may emit to a tine at angles to the "gamma photon" collision and attain direction determined by their vector collision angle. There is a chance that no gravitons would be emitted during this collision. It is unclear why the gamma photons would not appear to enter the gravitos to avoid each other. There could be a tipping point at which gamma photons contain enough graviton density to appear to be of sufficient size to react to each other in a more nucleon like manner than photons. In other words they each act like they are the photon that they are, but in interlocution they appears to each other like a nucleon, and so they both act accordingly and collide!

Cosmic photons may also be formed by gamma photons being emitted at the same instant on parallel tines and at some time they absorb each other resulting in a cosmic photon traveling on a central parallel tine. These occurrences would seem to be limited to high energy photonic emission states such as near the surface of the Sun. e.g. solar plasmas.

A cosmic photon would appear to be the photon with the greatest graviton density of all photons. When cosmic photons collide in a similar manner as described for gamma photons. They may release all their gravitons into the photos (which is a lot of quantum energy for two particles to emit) along an almost infinite number of tines only to be absorbed on a rate averaged basis by other photons. This may explain solar gravitational anomalies. The cosmic photon is so graviton massive that when two are passing by a hydrogen proton they may instantaneously form another neutron on that nucleus and so form a tritium isotope*. This may suggest that the cosmic photon is around half the physical size of a neutron but unlike a neutron able to decay instantly to gravitons because it is not affected by incoming graviton effects upon colliding with another cosmic photon because of dimensional separation. Apart from electron positron collisions, Gamma photons can only be produced by larger atoms (or delta neutrons) with combined graviton matter caused by graviton collisions being enabled by the extraordinary neutrons usually associated with such atoms. I.e. gamma radiation)

Gamma photons like cosmic photons are unlike other light photons (including x-ray photons) in that regardless of the dimension that a proton is in, the gamma and cosmic photons can collide with or affect it. This would suggest that gamma and cosmic photons are in the photos and have a great affect on the propos but are not affected by it. Note: It remains unclear how graviton drag can (in vector sum addition) lead to graviton transfer from neutron to proton at the nucleon bond junction. This would seem to be the realm of further investigation for quantum physics, but I think it is perhaps the role of mesons. * This again suggests that even quarks and all other quantum particles consist in some way of gravitons as well as sub-particles of force (bosons).

The size of a gamma photon suggests that an electron is about a quarter the density of a neutron. This is not at all problematical because IT HAS NO MASS and being a fermion it does not exist in the gravitos dimension.

Photons of all descriptions can be concluded to exist as singles or multiples of a vibrating string or stream of gravitons (or quanti). Gamma rays and x-rays are high energy" (frequency) photons that are able to seriously impact atomic nuclei and lose energy via inelastic Compton scattering. They pass more readily through opaque and dense matter in a variable relationship dependant upon dimensional constraints and protonic vibrational states (temperature). This is similar to the way that visible light can be seen to pass through your cheeks when you put a torch in your mouth. The impacts on nucleons cause damage to atoms and so can cause damage to cells in living tissue.

Nucleon attractive force is subject to all manner of other forces at various temperature states, and protons can not always hold on to the neutron. Such an example is the OH1 isotope of hydrogen which has no neutron but a single proton. In nature the average behavior of atoms is all that can be said about atomic stability due to uncertainty principle (which doesn't just apply to electron motional and positional states). Atoms are fairly stable until the nucleus contains more than 83 protons.

The average distance between nucleons becomes so great that the strong nuclear force begins to lose its ability by (diluted magnetic and particularly gluon force interaction) to retain a strangle hold on the nucleons and radioactive decay of the nucleus is assured over time. This eventually creates new less massive atoms. The reason that the "heavier" elements actually exist at all can mostly be suspected to be by the original cosmic forces at work during the first instant of the creation. I do not accept that heavier elements are currently being created in stars, rather perhaps in the flares of black holes and perhaps supernovas. I would tend to agree with other physicists who have calculated that the "energy" no longer exists in the universe to allow any atomic element above iron 60 to be formed.

In the end it seems that I have presented my theory along with my speculation. This goes to show that speculation, hypothesis and theory are the same if they can show answers to questions raised by observance and still remain faithful to unambiguous scientific law.

So what is the potential effect on our home called Earth? And are there any problematical energy related effects made available by this theory?

Fortunately the region of the universe around us seems to currently have fairly stable GD and the best "thermometer" for observing the state of the universe is our Sun. While the temperature of the Sun remains stable we can know two things. 1/ It has not run out of fuel. 2/ The GD has not changed. The other parameter apart from the previously mentioned energy divergence loop that helps keep the universe in GD equilibrium is the fact that cmf is proportional to the distance across the universe less the combined size of the number of gravitons on any particular gravitine. So areas with greater GD such as the regions at the centre of galaxies, being of extremely high GD density (The cause of black holes) have gravitons emitted at more rapidly decelerated velocities and hence less force/energy than areas of lower GD such as in our solar system.

Being a star on the small size it is likely that the Sun therefore allows graviton transitions and emissions at very high velocity and the average backfill of GS is greater here than that surrounding a black hole. This difference sees a magnified GS affect on the black hole but keeps our Sun's gravitational effect a little less than it would be in comparison.

There is no free energy available by this theory. It has either already been thought of or is already in use. (i.e. nuclear, solar, wind and "Earth temperature differential over depth" technologies). Nuclear fusion is not (by this theory) even able to be self sustainable on Earth let alone make power. The reason that scientists have embarked on such a very expensive, time consuming and most likely futile effort in this direction was because they mistakenly assumed that solar fusion is self sustaining due to temperature or magnetism or some such. They did not have a clue that solar nuclear fusion could be caused by graviton transitions through nuclei within the Sun.

What would happen to us if the GD decreased for some reason?

The first thing we would notice is that we would feel lighter and our "half kilo steak" would weigh less. Unless the change was drastic the planetary rotations etc would not change much because the GD change would affect the whole solar system en masse, as well as the effects of motion coincidently and retroactively. We would however notice the Sun lose temperature. If the converse were to occur the opposite effects would be observed and perhaps the planet Jupiter might sometime in the future begin fusion and start to "shine".

The effect with the greatest importance for science at the moment would be an observed change in "c". (Unfortunately and I hope not portentously, the fact that the Sun is cooling is already a worrying observation)! Significantly: scientists reflecting laser light from the Moon have observed a change in the time it takes for the return trip for the light. They now have three terms to choose from to decide which one/s must have changed, distance; time; or the speed of light. Of course the "c" constantists will conclude that the distance between the Earth and the Moon has increased.

There is no known cause for this to occur, because all changes in motion requires (guess what?) That's right; a FORCE. So until they find the alien "ufo" responsible, I'll settle for a "c change". Therefore I conclude that we have the first indication that light is slowing down. If we continue to measure time on Earth with cesium clocks, or if we were even stupid enough to use light to measure itself by its own wavelength/s, (duh!) we will of course measure no difference then, and the town crier will continue his indolent pacification while troops are surrounding the city. "It's nine o'clock and all is well".

THE FORMATION OF THE EARTH:

More speculation; this time about the formation of the Earth during the first short time frame, and continuing for possibly centuries after its creation:

The Earth would need to have been formed in area of dense cosmic matter within the universe, small enough to avoid being turned into a neutron star or magnetar, yet dense enough to form a wide range of low and high density atoms.

When the gravitons arrived, the agglomeration of atoms and molecules so formed was forced together in the form of a sphere which mirrored the extreme forces of gravity now acting on it by shrinking into a dense ball. The Earth began to orbit around a huge collection of lighter atoms that gathered together at a slower rate (because of the distances involved), which finally became the Sun a short time later.

The stars at other places were also beginning to shine and their own light (now having severely slowed down), was on its way to soon be seen by a theoretical observer on Earth.

The Earth became very hot at first. Elements and molecules as well as larger agglomerations of matter fused together. With a now rotating Earth the heavier elements were spiraled and mixed as they traveled towards the outer crust of the Earth. The Earth was hotter towards the outside than the inside due to reducing graviton "energy" loss with depth. The pressure inside the Earth due to expansion of matter was very great and it remained that way for perhaps decades or longer because the initial graviton assault was very short lived and decayed rapidly.

Water was formed in the first instance, from which oxygen would later be produced. The Earth's crust cooled by BBR at high velocity, but on the inside heat was convecting in towards the centre from the molten mantle. and at some stage, as the initial graviton assault subsided and the pressure relented to a point at which superheated water (held as liquid under high pressure) turned to steam, the Earth rapidly expanded to about twice it size. The continents were violently separated and super heated steam and magma was ejected in vast quantities into the new atmosphere thus contributing to it, specifically with nitrogen but (more importantly for our existence) with oxygen.

As the Earth continued to expand at a vastly reduced rate and the atmosphere slowly became less dense it would have at some time rained "oceans of water", thus cooling the tumultuous volcanic rents now seen to be scarring the crust on the continents and finally forming the oceans. Other large chunks of projectile matter with lots of momentum were accelerated with such explosive force that they reached escape velocity and recombined to form the Moon which began to orbit the Earth and rather strangely spin in such a manner that one face always presents towards the Earth. Yoda says "strange is that".

As the Earth expanded, its rate of rotation would also have slowed by conservation of angular momentum, but the molten and far less pressurized molten core which remained, would still continue to rotate with inertial velocity which has gradually decreased by friction until we come to the present time. This friction would contribute greatly to the Earth's inner core temperature being high enough to remain molten.

The magnetic field that the Earth now exhibits is probably less than in the past, and the volatile inner core could cause diamagnetic affects which could account for magnetic field strength changes in the past. Pole reversals have already been explained.

The amazing and progressive work being done by scientists and technologists at all levels should continue unabated. There is no need to "go back to school" because general theories of physics work. Perhaps a new generation of physicists could be sufficiently fascinated by the possibilities entertained herein, to focus on progressive science to help create hyper-technologies based on this or similar theories.

CHAPTER 16

PRAETOMS AND THEIR
TRANSFORMATION INTO ATOMS

A model of the cosmos can be theorized according to noticeable quark patterns in atoms which sometimes exist quite diversionably in the universe. While atoms may exhibit differing characteristics dependent upon energy states etc. The preexisting cosmos had no such problems.

One model I have developed is in an interconnected mass of quasicrystalline cosmic matter in dodecahedral meta periodic form, existing in multi-dimensional space.* this may have been in an arrangement of praetoms and anti-praetoms separated from each other with each quasi-symmetrical group of praetoms and anti praetoms sharing force particles. In this model the only particles are up and down quarks, North and south magnetic particles (magnetons) and gluons, all constitutive of the ingredients of gravitons. *(logic decrees that this arrangement must continue infinitely. But as I previously stated you can't always trust logic especially if the cosmos is a multidimensional "universe").

The discovery of the existence of pentaquarks as well as such arrangements of atoms being able to exist in our own universe may lend this speculation some credibility.

Perfect symmetry of pairs can be achieved with five adjoining cosmic pairs, semi protruding from each face of a dodecahedron. An array of quads can also be surmised but it's not so pretty! (I have used a "quad" in explanations for the sake of simplicity. Calculating the arrangement and then rearrangement of pentagonal cosmic matter being converted to universal matter would be "no mean feat"). Note: A scientist; Yamamoto Akiji has developed downloadable sophisticated computer programs to enable study of metaperiodic quasi-crystals in three dimensions at least.

Gluons would have been the agent provided the bonding force equally with quarks and magnetons. (not with equal force).

Now if an up quark has a biracial charge of + 2/3 and a down quark is—1/3 then the praetom has a net balanced charge of +1 and the interlocking anti praetom has a net charge of—1. This is because the praetom "quad" can be seen to have; 12 up quarks and 6 down quarks while the opposite is the case for the anti-praetoms. So in the cosmos the number of up and down quarks and the number of magnetons of each pole are summative to an equal number of any kind, consistent with the up/down quark pairs. I.E: 24 Quarks per object and multiples of 16 magnetons.

Now the most amazing thing about this is that this means that each pair can be exactly broken up into 8 nucleons with three quarks each with an even number of opposite magnetons. That happens to be the exact number of quarks in a normal atomic nucleon. So it is possible for a "praetom pair" to form 4 protons and 4 neutrons.

From this you can also deduce that a praetom can be formed into protons with 2 up quarks and 1 down, while an anti praetom can be reduced to neutrons with 1 up quark and 2 downs.

*The idea of a multi dimensional cosmos aperiodic crystalline configuration is theorized by the existence of similar material in our own "backyard" which appears to exist in at least three dimensions as described by R.M. de Wolf, and van Aalst, in 1972.

I will also admit that the Cosmos may have consisted of cosmic beta neutrons in a perfect matter-antimatter zero sum arrangement and this will be presented at length. The previous postulation would declare that the universe is currently losing energy due to anti-matter "evaporation" so the further postulation is more feasible.

Regardless of the make up of the cosmos. Something happened that shattered its equilibrium, (which in human terms was on a massive and universal scale). At the instant of the event the first praetom/anti-praetom mass was shattered into irregular sized pieces, many perhaps even exhibiting grain displacement flaws and striations, and others displayed a total breakup to fundamental particles. A universally vast web of this pre-matter would have taken all manner of chaotic and informal shapes and densities.

It was only when these pre-matter clouds, pieces and even massive sized chunks collided (in the multidimensional manner I have previously described), did they lose energy, and in so doing found that their new and paradoxically hotter particles where being stressed by internal forces, and (as occurs in nature, (such as in a bubble) they mostly became round, and at the micro scale quickly became attracted together with bosons, quarks and nucleons etc. in their vicinity to become atoms. At the universal scale, black holes began to "suck up" everything they could at very high speed at first but very soon after (as per now) at a very much reduced rate.

Why some new pre-protons threw off a particle with a dipole and a net negative charge (electron) is unclear but I suspect it had something to do with the meta-periodic symmetry of the cosmic mass which caused charge asymmetry in the universal state by the breakdown of quarks to release neutrinos*. In which they could only be reestablished by such pre-protonic actions called decay. This is what caused the arrival of electrons on the scene. This fortunate happenstance enabled them to form bonds with other newly formed atoms. Unfortunately for neutrons their quark arrangement leads to total decay if they don't pair quickly with a proton. *By caveat: The proposed sub-particle of particles including quarks which allows color charge quanta.

In areas of space where the cosmic material; was shattered into Quark-gluon plasma, (which was the most common result foreseeable) when it cooled and recombined with gravitons to become a nucleon it was able to take an intermediate step to reform as atoms with only one nucleon and one electron. This occurred under the massive initial assault of gravitons and stupendous amounts of hydrogen atoms were formed and were thrust together by GD into massive stars. This occurred in as many places as there were stars in the universe at that time. There would have been plenty of free electrons around to attach to deuteron as well.

It is obvious that much of this new universal matter would have been so dilute that gravity "is" even now still bringing the clouds of hydrogen together to form new stars. However it is theorized that much of the energy/matter in the universe has already vanished into black holes and the cosmo/universe and we humans then, are living on divergence.

The neutron-proton imbalance in the universe requires that many sub particles formed through complete pre-proton degeneration to cosmic and gamma photons as well as simply forming protons with electrons. This is probably a direct result of neutrons being held captive in the gravitos simply because of its quark arrangement while the proton was affected by other dimensions at the extreme temperatures existing at the time. Even at this very moment stray bits and pieces of atoms are being streamed into the universe from stars etc. E.g. Solar wind.

CHAPTER 17

ATOMIC DNA

Now we come to the obvious question: what causes dimensional positioning of nucleons etc?

To answer that question, I am not exactly sure, but I postulate a description of atomic activity as follows.

The first thing to be noted is that quarks are "flattish" and triangular with the three color charge points.

Now different atoms have differing Hilbert space set* interference patterns that sometimes extend beyond the outer electron orbital. This results in many and various electron orbital and internal force arrangements and interactions. Different nucleons have frequency phase shift variations due to eigenvector changes in the orientation of their quarks. This remains common to any elemental atomic structure and is able to be transferred to other atomic objects for atomic bonding as the case may be. *I may be using Hilbert space as a loose term to describe a limited space with periodic and aperiodic wave sets which exist in a very small space. Crystalline elements and molecular structures would likely be periodic.

These eigenvector shifts also cause vibrating, strengthening and weakening of the internal force fields, sometimes even canceling them out by null node affects at orbital junctures. This results in the determinability of dimensional states within nucleons. (Particularly protons)

Neutrons are not affected by the charge-field dimension because it is charge neutral, and so their eigenstate positional possibilities may well limited to a single eigenvector shift capability in one direction. I.E. the quark Cartesian plane can either turn or tilt/flip but not both.

This does have an affect on the overall atom by (I suspect) setting the element melting and boiling points and density, by Eos gravitos and magnos effects as the neutron quark interacts with neighboring quarks. This is pressure related because pressure "squeezes" the atom and causes change in the inter-quark relationships.

Properties such as color, hardness, density etc are caused by the infinitely variable but quanta stepped eigenspace transformations available to the proton quarks, by charge-field, force-field and magnos affects. These properties are defined by particular orientation and vibration values at STP in relation to other near-field nucleons in the case of atoms with more than one nucleon. The H1 proton only directly affects itself.

Protons are then determined to be in or out of any given dimension by one factor; the eigenstate of this quark orientation, which could be related to the "spin" axis. This is one aspect of atomic DNA. This is all concluded regardless of the spatial eigenvector state of the whole atom or object. In actualizing the whole DNA the Eos also reads the neutron quark orientation (which as we saw before) also contributes to some properties because the neutron is in the Eos and gravitos dimension and not the charge-field. This is all being referenced to STP.

The Eos reads the eigenvector relationship between protons and neutrons, and this is seen as part of the DNA. Nucleon arrangement in a nucleus is stable and consistent with the element. Isotopic addition of extant neutrons doesn't upset the basic arrangement, and the properties of the element don't change significantly. The actual eigenspace position with respect to the proton neutron sets can also cause further variation of properties. This can be caused by the introduction of external charge and magnetic influences. Temperature and pressure changes cause variations also, by determining the number of charge particles within the quark, resulting in color transformations.

Now a quark is simply a core particle existing in a dimension () with the relative sub-particles having variance and variability with regard to both vibration frequency and orientation which is the cause of magnetic dipoles and charge signs. This causes the sets to be able to be aperiodic, which results in band gaps and Fermi levels in electron orbitals.

I should address the nature of the interference patterns within Hilbert sets which are caused by the variable frequency interactions of all of the nucleons within the nucleus. The "set" fields propagate instantaneously as they are only a "felt" effect, and it is the velocity of boson vibrations at "c" which causes the apparent propagation of the fields at "c".

Under magnetic and electric change influences, the charge-field, and force-field dimensional affects have little effect on the atomic properties because they cause insignificant though proportional effects on the eigenvector angles because of the far greater strength of the extreme and near-field charge forces which circumvent them.

The Hilbert space interference pattern sets between the magnetic and electric force fields are caused by the function of the nucleonic arrangements within the elemental atom, with limited ability to be affected by external influences at around STP. As I previously mentioned, the addition of "tacked on" isotopic neutrons has little effect as they are not bound and they are also unstable. But just remove or add one "bound" neutron whereby the interacting nucleon arrangement is changed, and you will observe drastic changes in the properties of an atom.

However it is assumed with good reason that even though the neutron quark only exists in one extra dimension*, () it is still affected by extreme near-field affects which gives it the ability to interact with proton quarks and thus cause them to dimensionally shape shift. This is also further enhanced by gluon positioning which is in turn determined by the atomic structure. The properties of the elements or molecules are due to the direct tension of the combined eigenstate values of all of the proton quarks in any given nucleus.*All nuclei have protons which must exist in the first four dimensions as well as the Eos, magnos and electros. Other dimensional shifts are determined by functions as already stated. Any nucleus without a neutron is condemned to include the photos as its only other extra-dimensional state capability. Quark—antiquark dimensional status will be analyzed shortly.

Now we can conclude that the complete atomic DNA consists of interference or harmonic near-field patterns by proton and to a lesser extent quark eigenvector states and vibrational orientation, amplitude, frequency, and phase relationships at the near-field and extreme near-field level. Nothing in an atom occurs in a linear fashion. It always occurs with instantaneous steps which are causative of quanta and sub quanta integer steps.

This all seems to have left the poor little ol' electron off the hook as far as the DNA goes*. This is actually true because the electron is only reactive to internal and external; atomic influences. * At STP

I must admit that with multidimensional behavior, much work needs to be done in analyzing Pauli Exclusion Principle extensions and preclusions within dimensional arrangements, in that; particles are able to occupy the same space time and may be excluded from any Pauli principle determinations between dimensions. I am not "going there" in this analysis!

In concluding further; I would suggest that charge field and magnos dynamic interference sets, as well as gluon arrangements are responsible for crystalline and magnetic domain structure, elasticity, plasticity, hardness, and every other property apart from changes introduced by the external environment, or also isotopic and ionization affects.

NUCLEON DYNAMIC STRUCTURE

A quark is envisioned to be a triangular flattish planar object with a small but real three dimensional shape I.E. a flat polygon. The quark exists simultaneously within a "dimensionally shifted" magnetic dipole "containment" structure. The dipole/s is able to move spatially in any orientation within a nucleon without causing any affect on the eigenvector or value of the quark plane even though occupying the same space time. It also has no effect on the gluons which are in another dimension/s. The neutron quark eigenenstate is thought to be static because it has nil resultant charge value and would in that case not exist in the charge field dimension. It does however contain a normal dipole/s in the magnos.

Hilbertstate interference sets can extend beyond the outer orbitals in certain atoms as the case may be. This can make them very reactive as in the case of the oxygen atom, or totally inert if the "set" terminates at or within the outer orbital. This feature along with electron filling, results in the capacity or not for chemical bonding of atoms. This process may be further affected by phase relationships.

Quarks and proton quarks in particular, can be oriented in any direction with quantized color transformability. They can even rotate around their central axis a full 360 degrees, and my even spin/vibrate. Thus spin may well be a function of particle vibrations in some cases. It must be remembered that in atoms, electrons and quark "spin" is not transferable to the magnetic dipole or any other dimensionally diverse particles and visa versa. This is all caused by interference pattern force effects in the extreme near-field and in the intrinsic dimension/s.

Quark color transformations are caused by the addition or subtraction of individual bosons which account for the integer stepped behavior attributed to color states.

I have previously described the God code which covers several physical parameters. This is obviously part of a larger code which if you analyze the dimensional capabilities beyond the first four, you may see the possibly of an eight bit digital code with respect to other fundamental atomic characteristics.

We all recognize the elemental symmetry within atoms. We can now conclude that this is because of the periodic or aperiodic symmetry of the interacting Hilbert space sets of the dynamic charge and magnetic fields as well as gluon extreme near-field bonding force effects.

An alpha particle can cause damage to atoms in human tissue. This is basically caused by the fact that it has no electron orbitals and I theorize it to exude a Hilbert set interference pattern which is 180 degrees out of phase. This effectively annuls the shielding effect of the "tissue" atoms orbitals and it goes right on through without even collecting any electrons even though it might knock some around a bit! It is easily able then, to reach the atomic nucleus and because of its nucleonic interference it is able to fundamentally upset the properties of the atom, which in turn affects its atomic bonds which destroy molecular arrangements in a cell. It is free to roam from atom to atom and so wreaking havoc. This could also explain its supposed ability to tunnel out of the mother atom.

If a way could be found to (re-phase) them back to helium phase status, they would attract electrons and turn back into a helium atom and consequently become inert. It may be that the nucleon bond arrangement has physically been distorted in relation to a helium atom

You can also see how high velocity "neutral" neutrons and alpha particles can travel right through the electron orbitals and drive a "wedge" into a nucleus and split it and this is the simplistic explanation of the cause of fission.

In order for scientists to be able to fulfill the alchemist's dream of element conversion, they would have to (apart from gathering the required chemical building blocks) need a near-field "electro—bond force" radiation ad not an emr. Unfortunately up to the present, and at STP there is no know way to derive such radiation.

We see this in the fusing of hydrogen to helium. However this only occurs at very high temperatures and/or pressures. If we could strip all of the electrons from a very large nucleus we could have a "tool" to work with.

To strip electrons, it requires a force large enough to change the Hilbert set phase by changing the nucleon arrangement (we would probably consider such an object as being very unstable). The other problem is that to get at the nucleus we need to strip the electrons. This may sound like a great dilemma, but it may be possible at extremes of temperature while the Eos is "on vacation".

If this phase change occurs in alpha articles, then we might well ask; why not in whole atoms? The problem of course is how to then keep the nuclei and the matter object together without electrons. This may someday be possible with superconductor magnetic fields, yet currently at the extremely high temperatures required to perform the task, such a process is going to require some serious insulation!

If such material can be created in the laboratory, we may then have a tool to enable room temperature fusion to occur. However confinement of such material and temperature and process control appears to be currently very problematical. Note I didn't use the word impossible!

Can you imagine a lump of matter without electrons? (Not neutronium) Well imagine away, because contrary to the popular saying, I say that it is imagination and not necessity which is the true mother of all inventions.

I may be totally or partially incorrect in my theorizing, (which is up to you to judge and correct even). However this example shows that by delving deeper into the mechanics of atoms, imaginations can be simulated, and if there is any outcome of positive value to be realized; it is exactly that.

In the end we can conclude that relativity is "messing" with time, while multidimensionalism is "messing" with space. Both can be utilized in an attempt to arrive at meaning, and rather than doggedly adhering to just one model, let either be utilized as necessary. However if the case for the superiority of a particular theory arises because of empirical scientific method being applied, may the model which is a better fit to observations be staunchly defended to the possibility of the dis-accreditation of the other as not true.

GRAVITON FRICTION:

Considering that we should already have learned enough about the process of boson mechanics by now, we can address this subject, but first we will qualify some necessary attributes of gravitons and bosons.

1/ A graviton can consist of any number of basic gravions up to an as yet unknown limit. We can call that limit graviton charge Gc.

2/ A photon or radion-magneton stream particle (ramaton) carries fully charged gravitons only.

3/ Gravitons and bosons may occupy the same space time, either directly in the case of bosons of different kind or by dimensional allowances.

4/ Gravitons and individual bosons are passed to other nucleons for parity purposes by the mediation of mesons and quarks. Even though this requires the motion of force particles, the overall reactions within the AMO is randomly stable with vector "sum zero" affect.

5/ "Every action has an equal and opposite reaction" is a law that particularly applies to force interactions, and therefore it applies equally to force particles such as bosons and gravitons while existing in the dimension of the gravitos.

This is the reason that gravitons (even though able to pass through each other with no physical effect), can have a reactive, velocity reducing and course changing affect for the following reasons.

All bosons being force particles of three basic types have interactive force attractiveness and or repulsion as the case may be. Gravitons passing through each other are almost always trading bosons whenever there is a greater than two boson or quantum difference in their G charges. This difference would reach equilibrium in a closed system. However the system is far from closed, with boson and graviton transfers occurring in every AMO. Also gravitons are forever being emitted by photon/photon collisions.

When gravitons "collide" and a boson transfer occurs, there will necessarily be a reaction because an action has occurred whereby a boson or two was absorbed into another transiting graviton. The velocity and directional change will be "inverse energy loss proportional" to the direction of motion of the transfer. This can result in an eigenvector plane transformation of the directional change moment. It is important to understand that "no graviton will ever be more than two quanta different than any other graviton in the universe". They will always strive for parity. In that case two gravitons colliding could seem to swap just one boson such that the net result is the same, but then we have the conclusion that gravitons simply passed a boson between them simply because they could! I would suggest that in such a case no transfer or energy loss occurs. Graviton transitions through the same space time are allowed by the function of a gluon.

When a graviton transits nucleons, it may or may not transfer bosons dependent upon there being a disparity of greater than two within any given nucleon or if the graviton has a surplus of quanta. This will be more often the case than not, and visa versa. If a graviton travels long enough through an AMO it may then become "G-charge" depleted at minus three bosons and cease to exist by total quantum absorption into the nucleon, because of the above law.

So the friction which a graviton possesses is not a friction similar in any way to that which is described by classical physics. In fact a graviton transitioning a nucleon could "care less" about the number of bosons in the nucleon. The only thing that incites the friction to occur is G-charge charge disparity. The graviton vibration amplitude remains unchanged because its "energy" charge value is numerically related to the quantity of sub-bosons it contains.

A graviton would become rapidly depleted at the event horizon of a black hole because of the extreme G-charge disparities it will find there.

The reason that photons and ramatons can be emitted from an AMO without causing any change in its "mass" is because they are massless objects.

We should now turn our attention to photons and how being massless, they curiously appear to have momentum which does se to cause inertial effects upon striking AMOs. This is because they contain gravitons which (once released into nucleons) act in the same manner upon transiting those nucleons as we have just determined. This qualifies my prior explanation of the Crookes radiometer.

Heat is caused by graviton transitions through nucleons because the energy loss in gravitons is caused by the boson loss to the nucleon, (at some stage in "momentary" time) it is able to transfer out gravitons or bosonic matter to other nucleons and they vibrate at a greater amplitude, and in the interest of environmental energy parity the end result of this is either the emission of BBR or the attainment of the quanta step to creating at least an infrared photon which will be emitted upon reaching the surface of the AMO by whatever means available. Beulah! We have heat.

THE STARTLING DISCOVERY OF HOW A PARTICULAR MATTER AND ANT-MATTER COMBINATION ALLOWS THE FORMATION OF AN ELECTRON AND PROTON OUT OF A BETA-NEUTRON*, RESULTING IN THE CREATION OF A H1 HYDROGEN ATOM. This discovery also shows the mechanical makeup of the various parts of atoms.

Discover the perfect beta decay; positron neutron and electron decay into a gamma photon etc and the amazing conclusion that an electron mostly consists of antimatter! * An "alpha neutron" would be the normally theorized "udd" neutron.

QUANTUM BIRACIAL BALANCE THEORY:

I choose to disagree with Feynman. "Nature is not absurd!"

I can see why he must have thought so, because how you can get a—1/3 charged down quark to decay into a +2/3 up quark and end up with a—1 electron and a neutral anti-neutrino is beyond me and (I suspect) classical math!

Modern quantum physics is in a pickle: The sums don't add up. More massive particles turn out to have excruciatingly small actual mass. The whole "discipline" is squirming around a quagmire of inconsistency. Mass seems to not actually be mass, and energy is not really energy. Virtual and disappearing particles abound and some just appear out of the "vacuum"! Specific gravity and atomic mass relationships demonstrate wide anomalies which I intend to bring back into the fold of reason.

When faced with these dilemmas; instead of questioning the starting point the physicists are acting like a baseball team without a bat, wondering why they are always scoring run-less innings! I'll present a picture of an atom and especially an electron which seems to be the latest high tech "bat". The electron can even remain in orbitals for reasons other than "magic", and it is even able to obey all the classical laws of physics in so doing!

This doesn't mean that quantum mechanics is not weird. It Is! But regardless, logic declares that it should still obey the known laws of physics whenever its actions demand it.

This intrepid yet tremulous attempt at enlightenment also shows what a neutrino is, and what it does; also the role of bosons in forming dipoles and charge poles in electrons and more.

If you bear with me, you should see this all come together in a congruent "physics obeying" whole.

A charge neutral magnetic pole (if one existed) would consist of four magnetons positioned in a square arrangement of cross linked anti-magnetons in one cross line of the "x" and two magnetons on the other, with (weak force) +w bosons and (anti)—w bosons providing the bonding force to hold them together. The magnetons are consistent with causing a north pole for example and the anti-magnetons would make for the south pole.

From this we can declare that without antimatter there would be no observable charge signs or magnetic dipoles in the universe*. The idea of the observability of the existence of a single point charge or a single magnetic pole is an absurdity. *In fact I intend to show that without a large quantity of anti-matter we would have no universe and that it is actually biracial attraction BA which is the basis of "force" and hence energy.

These biracial magnetons actually become arranged as a "dipole" within a nucleon or other particle in a straight line as single or multiple dipoles with two magnetons at one end and two anti-magnetons at the other, being held together by the said bosons and anti-bosons respectively. This same arrangement is carried over to an electron as we shall see.

Before we get to beta decay and its variants, I must show a simplified diagram of a cosmic or beta-neutron (which is probably the original praetom antipraetom pair, previously described and which has become the cosmo-universal), proton and electron and all of their component parts in order to be able to explain how beta decay can occur with just one anti-neutrino left over which is emitted at about "c" into the Eos, whereby over some vast distance it may be re scavenged or transformed by an unknown process.

First I will show a diagram* of a beta neutron (figure one) which exists, with half in the Eos and half in the gravitos. Contrary to common theory, this diagram shows that a beta-neutron consists of three mesons which are quark, anti-quark pairs which are bonded by gluons. The mesons themselves are bound to each other by gluons as well. *For explanatory purposes only. The particle proximity is likely to be close and multi-dimensional space-time sharing.

Separate from the mesons, you will notice a line of magnetic dipoles consisting of twin w bosons, and anti-w bosons, held together by neutrinos and anti-neutrinos as the case may be.

The anti-matter exists in the Eos and the matter in the gravitos. It can be calculated as shown that the neutron is charge neutral. The neutron is declared to be matter—antimatter biracial,* whether or not it will ever be observed by a microscope is doubtful. *This does not cause any change to its mass, however one can see that it has a quark bi-racial charge of zero, but more anti-matter than matter by just one anti-neutrino.

BA is also seen as the reason for atomic bonding forces. The races are

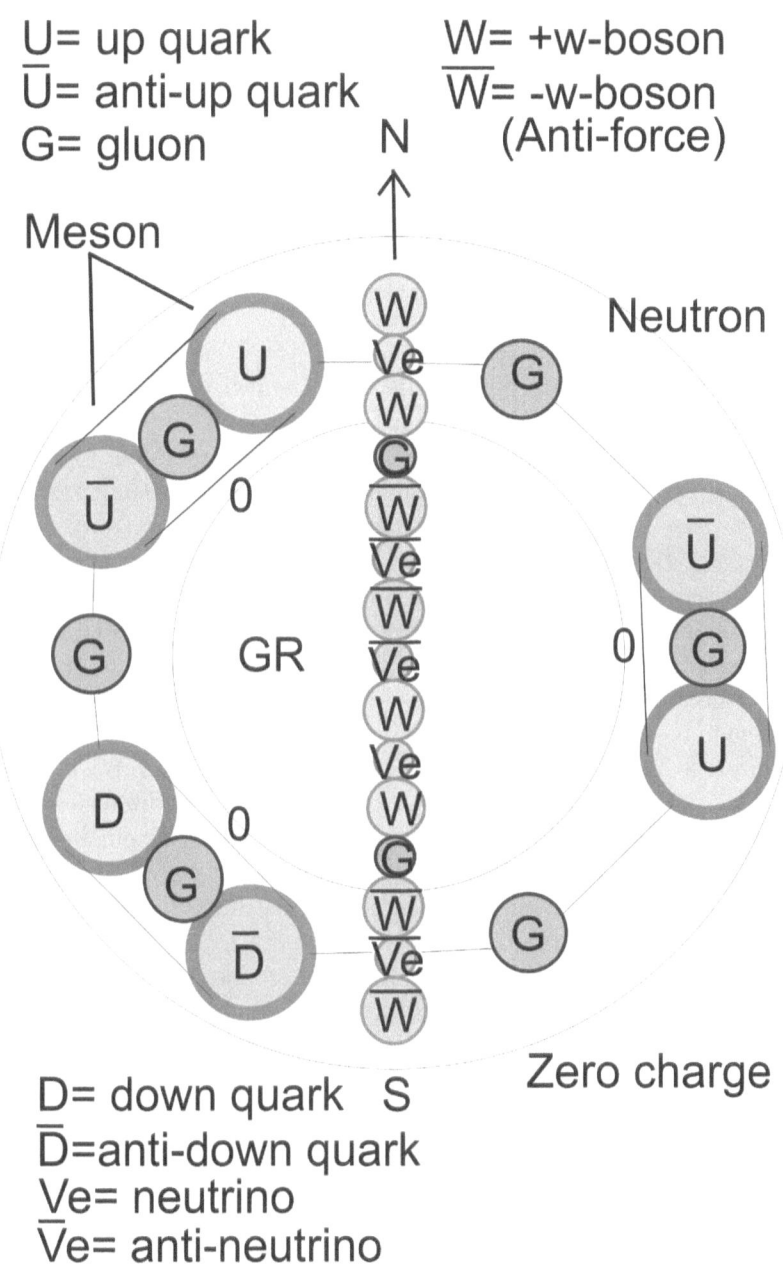

U= up quark
Ū= anti-up quark
G= gluon

W= +w-boson
W̄= -w-boson
(Anti-force)

Meson

Neutron

Zero charge

D= down quark   S
D̄=anti-down quark
Ve= neutrino
V̄e= anti-neutrino

FIGURE 1: charge zero, beta-neutron

Now if we study the diagrams of the proton and electron and calculate the combination of their parts you will find that they can almost totally be recombined to reform a beta-neutron". This would be by beta positive decay.

You will find that an anti-neutrino is emitted in the separation process because it is not required for the proton electron production* and if one is not available for beta positive decay recombination purposes by the process called "electron scavenging" then one will be gained from the graviton repository by a process called "bremsstrahlung gamma quantum". It is understood that massive relative force and energy is required for both of these processes to occur. *The baryon number remains conserved.

When a proton electron pair is created, by beta-decay two spare gluons are left over, but these are then available to be utilized by quarks in order to hold the nucleons together. Those two gluons are just sufficient that when shared with other nucleons they can bind a fairly large atom together. As the nucleus gets larger by the additional binding of other nucleons the two gluons become insufficient and the shared binding force in turn becomes less efficient because of the nature of the "space" filling arrangements of nucleons in the nucleus, and the atom cannot be arranged to maintain sufficient binding force to remain stable and alpha decay is inevitable.

You will notice that the proton in (figure two) looks similar to the standard "matter" model, except for the existence of two anti neutrinos. The following may seem to be an astounding observation but it must be realized by now that an electron is mostly made of antimatter and it exists in the Eos with parts in the magnos, and chargefield, with two gluons joining the dimensional parts together on the branes. It should be noticed (with equal astonishment) that an electron is actually the antiparticle of a proton, but only if you give back an anti-neutrino and re-dimensionalise it. (Figure three)

An even more confounding conclusion is that a "theoretical universe" such as this would consist of combined and proportionally equal biracial matter and antimatter constructs, with only a slight overall bias towards matter and obviously with the inclusion of biracial particles such as gluons, and (in a different way); positive and negative w-bosons. It could be perhaps concluded that the biracial sum at any point in time is zero and that matter is balanced by anti-matter! Shortly we will see why this is not the case in our cosmo-universe.

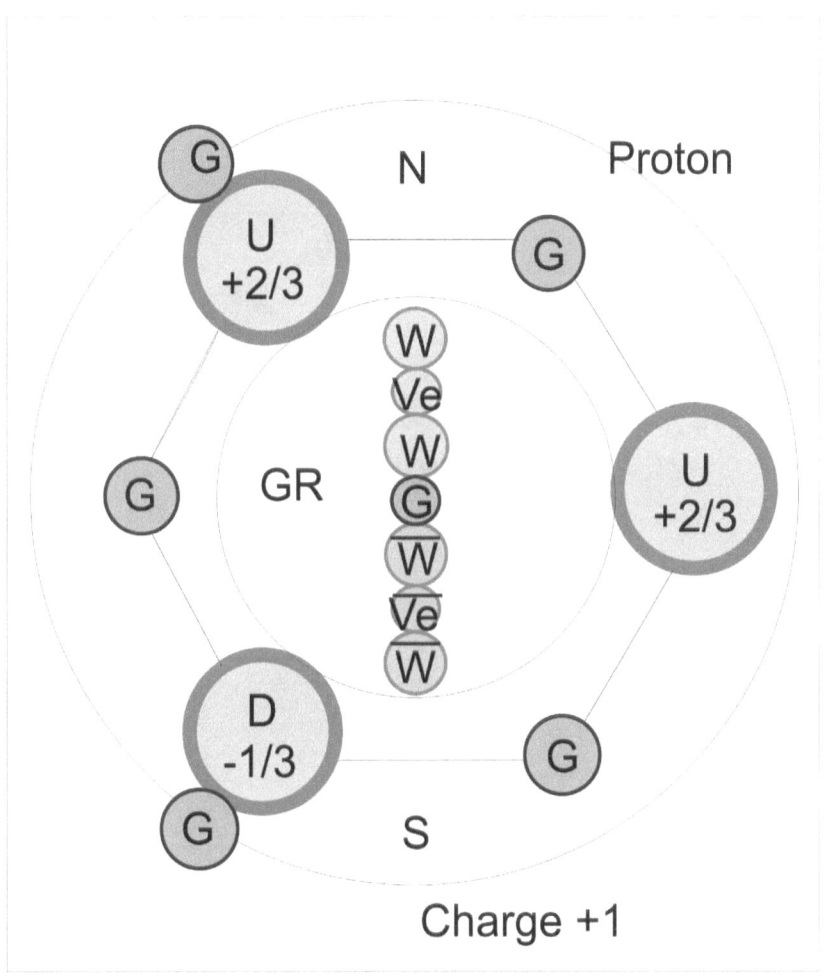

FIGURE 2

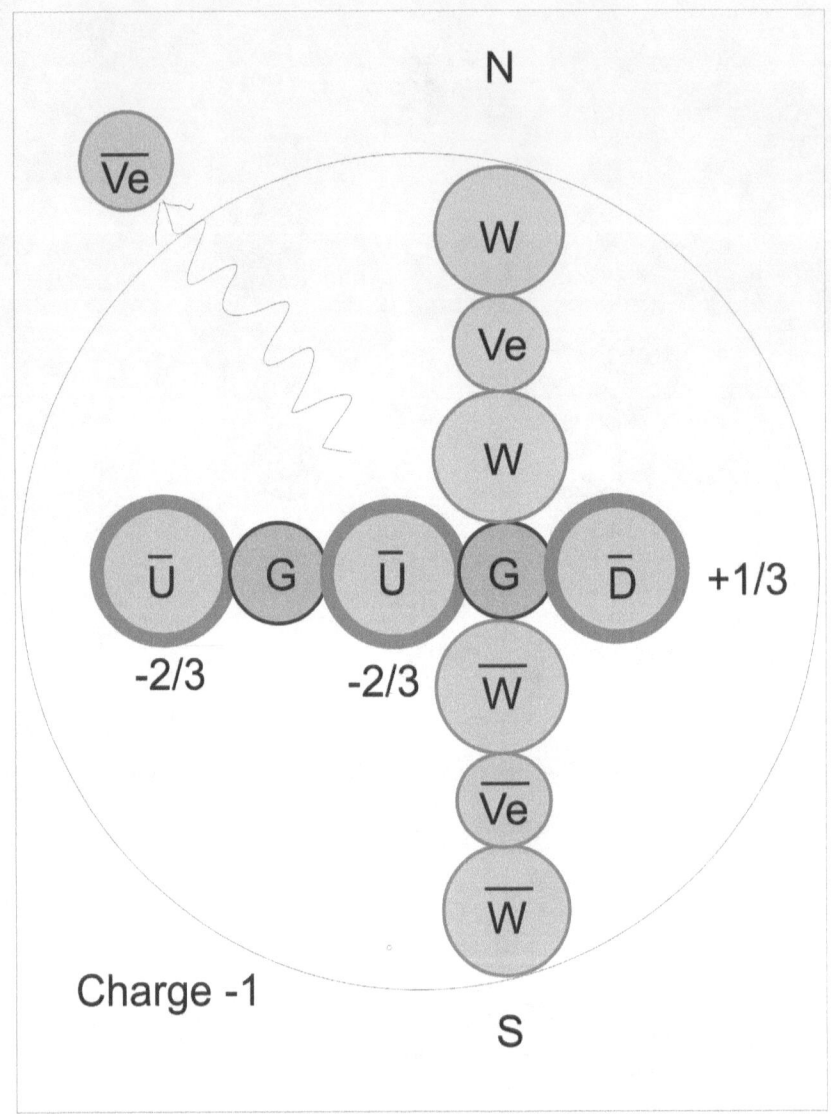

FIGURE 3

A graviton "G-charge" quantum, which is similar to a base level low infra-red photon, is likely to be configured as shown in (figure four). They both consist of a z (weak force) boson, a w boson biracial pair separated by a neutrino and a gluon. All of these sub particles are deemed to consist of smaller sub-sub-particles/anti-particles, which are able to transfer energy to atoms and photons etc. by force, which also enable color variation in quarks and anti-quarks.

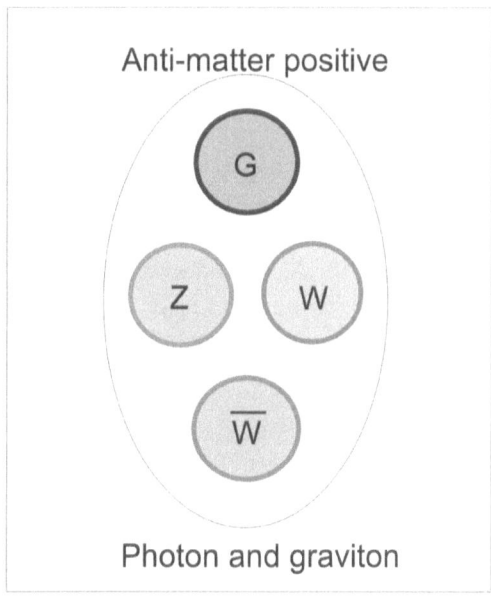

FIGURE 4: (one G-charge)

External energy states will cause the quarks to dip into and toss back into the "repository" as required. The energy state of the nucleon determines the value number of the repository. External energy arrives by photons and or is delivered by BBR as well as by graviton transitions through the nucleonic graviton which is confined to the Gravitos*. The repository is considered to only exist in the Eos (extra-dimensionally speaking). The particles could still be observed as probably a cloud surrounding the quarks etc. and dimensionally co-existent in space time as well. *I will now state an exclusion principle which is absolutely necessary if it is observed that matter doesn't lose mass or weight in any proportionality or dependence upon the energy state of such matter. The principle states; that "only one graviton quantum can exist in the gravitos inside of a nucleon at any one time and it must remain there until the nucleon becomes exhausted at the Bose Einstein condensate temperate. This means that the bulk of unused particles existing inside an atom (which are the quantum energy state gauge, exist outside of the gravitos.

Another principle states that "matter and an antimatter particle can exist together without annihilation if combined with a neutral gluon" in a dimensional separation arrangement. Even when so separated the biracial attractive force remains.

Also a gluon in the graviton repository is confined to the branes between those relative dimensions. It can never again exist just in the Eos to be able to form quarks, except at a black hole event horizon in quark-gluon plasma state. However it must be determined that at around STP any dual set of biracial particles below the level of a quark may remain combined by a gluon indefinitely and so undergo no decay upon "contact" with other such particles and or each other.

A gamma photon is likely to be according to (figure five) which is determined to be the case because of known electron decay mechanics. It turns out to actually be a dimensionally shifted magnetic pole. If two (which is the quantity emitted by an electron during its decay) were to combine to form a magneton or dipole, they would need a neutrino or anti-neutrino for the process to occur. Again, one could be scavenged from the repository if required.

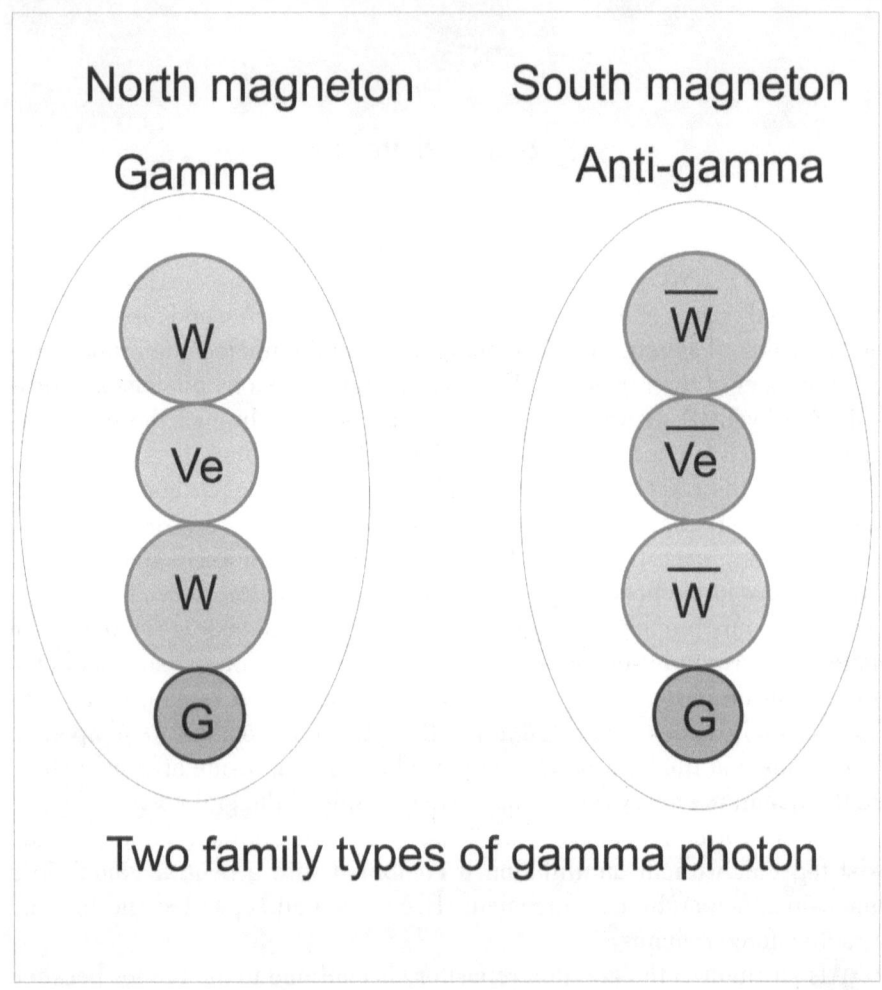

FIGURE 5:

If a gamma photon construct existed in the magnos it would be a magnetic dipole! Or in other words; it would be a universal magneton.

When an electron decays into two gamma photons the gluons allow the anti-quarks to be absorbed back into the Eos and the new photons are translated by the gluons into the photos. In normal atomic circumstances the matter gamma is ejected from the atom and the anti gamma is attracted into the proton by racial charge attraction. Because gamma photons have no quarks they can have no electric charge.

There appears to be some violation of CP symmetry here. However this is not unexpected because observed CP invariance is already thought to be caused by the disproportionality of matter-antimatter symmetry in the universe. Quantum symmetry is not damaged because quarks are determine to be the cause of hadron quantum numbers, and this theoreticist is not convinced that color quantum is a particularly necessary theory for strong force,* otherwise one might expect that the force would grow proportionally stronger with temperature. In that case please explain "quark gluon plasma" and rapid beta decay at extremely high temperatures. There is no law which suggests the conservation of matter-antimatter parity! *If color charge (whatever that may be) is responsible for the strong force then I say that it is a function of the Eos because it seems to lose its strength outside the "Eos effectiveness temperature window". Observe the behavior of nucleons at both extremes of temperature. It is likely that color charge is the quantum control for the whole nucleon energy level state within the window.

A "cosmic ray" photon is likely to be similar to a proton, but without a graviton repository. It would therefore have no mass and would be a bogus-photon also existing and traveling in the photos. A radion would likely be as per figure 6.

The arrangements of leptons may be realized by similar modeling techniques. The strange lepton and neutrino combinations sometimes observed are random anomalous occurrences and some of the particles are extremely short lived. However they do give further insight into the structure of the atom. E.g. a pion is basically a neutron with an extra magnetic dipole, but it decays very rapidly by electro/magnetic entropy. This model does not damage conservation of lepton number.

At quantum levels size and spatial position have no relevance even though spatial orientation is important. Other status symbols that are relevant are race, family and dimensional states. I.e. it can be stated that; "A photon is a four sub-boson biracial family group, in the Eos and the Photos and Forcefield dimensions". A graviton is similar but dimensionally separate as it exists only in the Eos and the Gravitos, and they are therefore able to pass right through each other without affect.

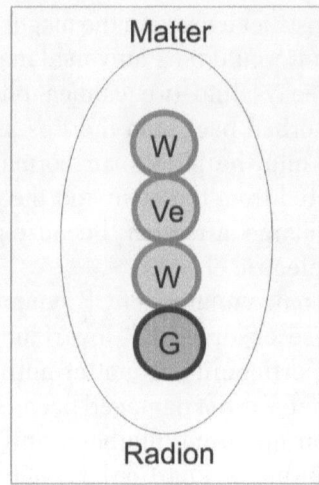

FIGURE 6

There is another force which we can now analyze and that is the atomic separation force. We all know that atoms can "push" on each other, and they tend to keep general separation when not chemically bonded. From this we can determine that there is some repulsive force other than binding and chemical bond force that is at work here.

This can be explained by analyzing the electron model in figure three. Upon a close examination you will notice a strange thing. The outward facing anti-down quark has a near field charge of +1/3 which in that near field sense shows a net omni-racial repulsive charge to any adjacent atom which will be responding in kind. This repulsion which I call "like matter repulsion", is powerful enough to offer strong resistance to the natural cohabitation of many atoms which is usually proportional to the number of electron in the outer orbitals.

This force in combination with nucleon quark forces, as well as the chemical or valence bond force may well be responsible for some elemental properties.

A neutral pion plus two biracial charge pions is by this theory a delta neutron which has four mesons and still only a neutral charge. It is only thought to have a single dipole because just a single gamma particle is ejected during the process of its decay into a positron (proton) and electron. A down quark anti-pair meson must also be produced. Whether a meson can be observed outside of an atom is debatable but not thought probable by large consensus.

FORCE CONVERSION TO ENERGY at the bosonic level.

Now all the graviton and bosons in the "family" groups within the graviton repository within a nucleon, all contain "like matter repulsion" content and all their efforts at keeping their distance from each other, results in a shambles of super-velocity vibrational motions. This affects the mesons, quarks and electrons, and they then in turn cause the atom to exhibit a certain temperature vibration.

The more packed in the bosons are, the greater the chaos, and the atom tries to expand by "vibrational matter expansion" not because they have momentum but because they exist in the Chronos and so exhibit time delayed decision and reaction processes, and they begin to attain closer proximity behavioral effects which causes them to behave more elastically.

The space required for the motion to occur in is attempted to be increased by force unless other atom's opposing forces are in near field effective range and the temperature will then increase further.

This mostly occurs by the force imbalance caused by biracial imbalance which in turn results in a force which is (in vector sum) attempting to enforce "particle family quanta" expulsion from the nucleon. Parity occurs when the vector sum of force within an atom is equal to within one quantum of neighboring atoms, or the Eos as the case may be.

When the atom has an avenue to a meson or quark which is vibrating at lesser amplitude, its nodal force sets will be protruding further into that adjacent nucleon than visa versa, and it will send streams of gravi-photons and or other photons into the lower "energy" level meson or quark until close quantum parity is reached.

If the atom is at the matter event horizon it will emit BBR and bosons to tines, gravitines, and into the propos; streams of photons etc. until the local universal parity is reached and the Eos says "stop in the name of the law!"

An atom at a temperature just prior to the BST will be at maximum boson capacity, and remain that way by conversion to a new praetom at zero degrees k in the cosmos state However as we have seen, if the atom drops in temperature to the Bose—Einstein condensate level it may contain one or nil bosons (graviton). In that case it would not be subject to gravity, only to BBR and meson quark energy transference as well as some slight quark/meson force interference at (I suspect) very low frequencies.

Now by analyzing this process we can see that force is from particles which react to energy transfer of more force bosons which causes them to vibrate and move around more as stated, which causes a rise in temperature, and which also causes them to keep as close to parity with the neighbors.

The force particle transfer, or spatial directional motion of bosons is seen as kinetic or a "release of" energy, while the vibrational state and quantum value of the particles in the repository is the rest or potential state energy value as the case may be.

Now this may look like a fancy "model": If I could only fuel it up and turn the key! I guess that would be a bit premature, because at the moment all we appear to have is a box on wheels. At least I reckon it's beginning to look the part now, and we may even have discovered a "freeway" for it to travel on by this whole theory. I wonder how "tricked up" it will turn out if real physicists put it in the shop and "pimp my ride". Enough of car metaphors already! OK

I must admit that I have given fusion energy research a bit of a "bagging" which may not have seemed fair. However by now you might be beginning to see things my way, yet it must be pointed out that ALL research is necessary and helpful, even if it may never meet with the hopeful ends. The scientists so involved are of a sincere and "big bore" caliber*, and if I can be allowed to pick which technologies are going to lead the way forward, nuclear fusion is up there with the leaders. I propose that the other hopefuls are laser, definitely synchrotron and super-conductance, as well as the old faithfuls of electrostatic and magnetic physics. If I could only get my hands on a bottle full of gluons, I'd see you all around Andromeda! * This is no fake humility on my part when I state that compared to such physicists I am in no way able to be a peer. I have simply stumbled upon an idea which I have attempted to explain with all of my obvious failings in math and physics.

To discus the processes which cause both positive and beta negative decay; we need go no further than the flare of a black hole. At the point near the singularity, the temperature of the flare will be very near BST and apart from much nucleon separation into quark gluon plasma etc; many neutrons will only undergo very rapid beta negative decay. This will be occurring by the extreme temperature vibrating the neutrons so violently that the gluons in the mesons lose their grip simultaneously and the biracial parts of the atom bifurcate and reform in their relative dimensions. Both types of decay will be occurring together as we travel out along the flare.

Further out along the flare as it continues to cool the neutrons reach a temperature whereby they still have sufficient energy for neutron recombination via positive beta decay and they "mate" with proton-electron decay sets. Masses of similar atoms are likely to aggregate and form clouds of elemental gas which then cools into liquid and then a solid mass.

The process of the formation of elemental matter in aggregate masses of similar atoms can be fairly well imagined. The real difficulty with the formation of matter is with the creation of the similar atoms themselves, and more especially the larger ones. The logic must admit that the process must occur under some form of control, because only half the neutrons of the original cosmic matter can beta decay during the process, and they must all do it simultaneously and eject their electrons and combine with the neutrons that "decided" not to decay. This by reason is a profoundly enigmatic and sobering process which is actually not even to be logically entertained as a possibility because of all the collisions and opposing positive and beta negative decays going on throughout the self defeating struggle for the hopeful birth of such a nucleus.

This nucleus formation therefore can only really occur multi-dimensionally at temperatures which release the matter from the dimension of the eos. In that case the various protons and electrons can "pass right through each other" and it is the not necessary for there to be any miraculous decay similitude requirements.

CHAPTER 18

FOOD FOR FURTHER THOUGHT

A caveat is in place with regard to the following, In that current analysis of particle physics can be somewhat likened to the "observer" studying we humans fumbling with a Lego set and changing the shape of the object being formed at every turn. What I am going to present is perhaps going to pull a couple of pieces off and rearrange the object a little, and I'll also be adding a piece or two of my own. The object should still be recognizable and I hope we may see the blurry outlines of something cohesive and beautiful emerging.

If you haven't just "gone straight to the back of the book" you should by now understand how the following particles can be determined to be massless, but not matter-less. These particles except for energy particles are other dimensional, and lay outside of the possibility of direct energy matter or mass transmutability as previously described. However antimatter states are thought to be the transitional states whereby particles are enabled to pass through branes to other dimensions in which case other particles are formed by re-unification. Which in some cases (such as with neutrino-nucleon impacts) can turn a neutron into a proton and change one element into, say the next element on the periodic table, all within constraints of "charge current interactions". Particles may consist of two or more sub-particles.

The following is a list (not exhaustive) of subatomic particles and sub-particles including examples.

> "Energy" particles—graviton, photon, gravi-photon, radion ramatons and all quanta packets
> Charge particles—electron, beta particle, quark, neutrino boson
> Magnetic particles—electron, nucleon, magneton boson
> Strong force particles—gluon boson, magneton boson, quark
> Weak force particles—W boson, graviton, neutrino

Interlocution and boson transfer particle—meson
Bond force particle—electron
Separation force particle—electron
Antimatter particles—anti-particles of all types; these are particles
  I call dimensional transition particles that temporarily reside
  on the brane between dimensions or in other dimensions by the
  agency of gluons. Recombination with its antitype can result
  in other particles being formed in different dimensions. In the
  end all particles can be theoretically reduced to bosons and
  even below that as other sub-bosons and pure force particles.
  Neutrinos and antineutrinos have had no known function until
  this theory.

Note: The solar neutrino problem does not require neutrinos to have
mass, because the Sun (by my theory) should be cooler in the inside than the
outside. Internal solar convection streams will muddy up conclusions that
can be drawn, so the caveat applies. However it is possible that a neutrino
travels in the Eos dimension and because of the proportionally smaller effect
the Eos has inside denser objects and especially metals, then the neutrino
is (paradoxically) more likely to be reunited with nucleons in matter which
has space between the atoms, (such as the atmosphere) than the solid Earth
which it will more readily pass through. Even water will be more neutrino
opaque than the Earth.
The following list shows particles (and assume anti-particles) and the
probable dimension/s of their existence and influence.

Neutron—half in Eos, half in gravitos
Proton—Eos, multi-dimensional
Graviton—Eos, gravitos, force-field
Photon—Eos, photos, gravitos
Radion—Eos, propos, force-field
Ramaton—propos
Quanta—Eos, gravitos, force-field (Photons, Radions, gravitons
  and quanta are only dimensionally dissimilar).
Electron—Eos, charge-field, magnos
Beta particle—Eos, charge-field
Magneton, (W-boson)—Eos, magnos
Gluon—brane, trans-dimensional
Quark—Eos, force-field, charge-field, gravitos
Non graviton/force bosons—Eos, force-field, charge-field,
  magnos
Neutrinos—Eos and many other dimensions by gluon agency.

Meson—quark, anti-quark pair
Gravi-photon—Eos, gravitos

The felt forces of attraction and repulsion are likely to be limited to their own dimension.

It has been stated by others that there are 50 trillion neutrino transitions through the average human body every second. Although neutrinos cause no significant cellular damage, they may be a contributing factor in aging because of their ability under the before mentioned constraints to affect atomic structure.

It is theorized that graviton transitions are far in excess of 50 trillion and that the GD is so graviton dense that any attempts to calculate a number would probably be futile. Gravitons cause no damage to biological tissue whatsoever.

N.B. There appears to be three different quantum levels related to particle sizes: being photons etc, bosons and sub-bosons. There are also seven levels of matter, being; molecules, atoms, nucleons and electrons, photons, mesons and quarks, bosons, and finally sub-bosons. Gravitons and photons may be multiples of G-charge which is fully determined by the quantum energy state of the emitting particle.

THE ROLE OF BIRACIAL IMBALANCE IN ELECTROSTATIC CHARGE, MAGNETISM, FORCE, ENERGY TRANSFER and PHOTONIC/PROTONIC ATTRACTION:

As far as our biracial cosmo-universe is concerned, biracial balance is eagerly sought.

The hydrogen H1 atom is the most biracially volatile atom of all. Because if you strip away its electron it will be observed to have the full biracial imbalance charge of +1 which because it consist of only a single proton it then has the highest charge per unit matter density CPMD. This high charge centralization is also proportionally associated with it when bonded with other atoms. Hydrogen is seen to be a prime contributor to chemical induced electric charge processes.

The reason water exhibits some strange properties is because of the lopsided CPMD caused by the angles of bonding arrangement of hydrogen atoms in the water molecule. However I digress.

Within the atom (all atoms) quark and dipole positioning is caused by the biracial balance in a multidimensional framework. The dimensional character of the atoms is caused by the gluon positioning due to space filling parameters. This leads us to the amazing conclusion that the properties of matter are caused solely by space filling geometry, and thus the elements are like they are, and we've got what we've got without metaphysical reason!

If electric charge "v" is a physical reflection of biracial charge (no v sign) then it stands to reason that lightening is the net result of strong biracial imbalance BI, and it is actually a matter/anti-matter collision due to the initial ionization of the "arc" path causing sufficient elevation of temperature to banish the Eos "out the window" and the biracial collision can then occur. This is why lightening is strikingly powerful. (pun intended) Matter colliding with anti-matter is far more "energetic" than mere ionic transmission of charged particles.

I would consider that during the strike electrons are dimensionally shifted into the chargefield dimension because of nucleon filling distortions in the ionized atoms. This would cause beta plus decay into neutrons, which would rapidly recombine with protons and electrons to form hydrogen, ozone, tritium, and water molecules of various descriptions. The idea that electrical arcs may cause B+ decay should perhaps not go unnoticed.

If a proton is left alone in deep space, it will rapidly emit matter as "energy" in the form of BBR and photons, and because (biracially speaking) the proton is matter positive. It could also send out a positive gamma particle or two. It will devoid itself of all matter except perhaps the base level graviton until it reaches a close to zero k as it can get.

In the real world at STP other constraints prevent this occurrence and stable hydrogen ions are available to be the major cause of electrical charges because of their relative CPMD.

Extreme near-field to proximity-field BI is thought to be responsible for magnetism through the BI of w-bosons, and near-field to proximity field BI of quarks is responsible for electrostatic charge. External forces can affect these BIs resulting in motional changes and heat energy release.

Whenever a BI is of sufficient value to cause the displacement of a boson then a quantum value of force has been realized.

The photon/graviton to proton attractive force is mediated by the weak force z-bosons within them*. This is an extreme near-field force, but it can extend long distances from massive bodies with the combined effect of trillions and trillions of atoms. This can explain the "bending of light beams around universal bodies. *Z-bosons, like gluons are not "BI" they have an attraction to each other, but they remain dimensionally diverse. There must be a law similar to Pauli principle which states that no two z-bosons can exist in the same dimension within any quantum particle unless they are held separate by some unknown property of quarks or by gluons. Their force affect is trans-dimensional.

ENERGY, MASS, SPECIFIC GRAVITY AND SPECIFIC HEAT:

There is a relationship between all of these terms which is circular. It is this closed system of mathematical functions which gave credence to the idea of mass energy equivalence, culminating in $E = mc^2$.

If you attempt to calculate the volumetric weight of an element by the formula—Am=2An*Sg*Cg with related units, you end up with an apparent mass discrepancy which can be rectified (sort of) by determining that the electrons must have mass and by then adjusting the formula to what I call the energy/gravity factor GFe=AM*Sg*Cg you arrive at a correct relationship which is close to the actual realized weight of a particular volume of the element. This is not C(mol) which is the mass specific heat relationship rather than volumetric and you use the formula C(mol) = AM*Cg you have the mass specific heat relationship. The stretch to energy mass equivalence from there is (I must admit) very short and tantalizing.

However; apart from the molar specific heat anomalies which have already been addressed, there is another problem.

In the process of deriving molar specific heat we arrived at the term above that I called mass factor MF per atom. If you divide this by twice the atomic number, this then becomes the nucleon MF which should include a supposed mass component for each electron in the atom.

However upon analyzing the data it is very noticeable that nucleon MF for individual elements is very anomalous. Eg Al—2.38 . . . . Sn—1.67 . . . . Au—2.49 . . . . Cu—3.41 . . . . H—1.0013.

When you analyze this representative data set you notice that Hydrogen is the only one which seems to have nucleon mass plus electron mass per nucleon. E.g. gold has a lower MF than aluminum which is lower than copper as you would expect but tin is lower than them all. There must be something else going on here; but what?

If the overall atomic mass can be determined by nucleon mass plus electron mass but the individual nucleon mass plus one electron mass can't account for the mass we have a real problem.

The answer to this apparent dilemma is found when we relate these results to C and we arrive at the correct answers and the atomic mass results in the periodic table. You may by now be asking: If that doesn't point to mass energy equivalence what then does it lead to?

I contend that it only points to mass/specific heat equivalence and nothing else. If energy rises with temperature then we should see a proportional rise in the mass of the object. Of course we do not!

Now specific heat C is inversely proportional to temperature such that there is no real significant change in the mass.

Now we have the dilemma stated as: If gravitons are causing the mass by transitions per unit volume then what has that to do with specific heat and temperature relationship in such a manner? Everything! Firstly there is a flaw in the above formula for atomic mass in that it only works at STP. For it to work at all temperatures and pressures there needs to be a temperature and pressure related component in the formula. I am not going to attempt this myself. However it should be kept in mind with the following possible explanation.

With all else being stable; the specific heat of an object is temperature related as we have just noted. Temperature of objects (apart from any other external source) is caused by graviton transitions through the object per unit volume. If motion is caused by this, then the temperature will be affected by a quantity based on the proportionality to the energy of the motion so caused.

Now I would love to relate mass to Sg and to temperature and specific heat but if the number of graviton transitions per/sec is reduced, GD goes down and Er goes down which means that GD may be the thermometer which causes proportional change in universal average temperature.

So it is not actual temperature which has proportionality to C it is GD related temperature. Specific gravity has a problem as well

There is a need to bring C into the atomic mass calculations in a different manner than currently used for around STP requirements. The other problem with S.G. measurements is that for a start, science is comparing "apples" with "oranges" whenever it compares density between solids and a molecular liquid.

There is a second problem in that the water I am referring to is having it's own density evaluated at about four degrees c which gives it a particular C at that temperature. Comparing this with other objects which are not at or near their freezing points is going to see them have a different value of C. If this could be done, or better still if the density experiment could be done when other materials were at the same C temperature relationship, then there would be no mass discrepancy when compared with the H1 atom in a similar situation. The hydrogen atom is not affected by real mass discrepancy because it has no space filling or C variability with volume and little with pressure. Of course such comparison appears to be extraordinarily difficult to achieve experimentally.

This also means that other single atoms and molecules would have insignificant to no specific heat, because it would all be instantly sent to the Eos via BBR. This then leads to the conclusion that specific heat must be somehow relative to (not only density and pressure) but volumetric size and shape.

Of course anyone who has heated the tip of a needle or exploded some flour, would have realized this immediately, wouldn't they? If this is correct it has grave implications for the validity of the periodic table: Full stop.

Mass and specific heat can be vastly different "animals" then at the macro and micro level. This means that the current periodic table can only be correct for a particular and different state of material size, temperature and pressure for each element.

However given the computing power that we have at our disposal; a three or four dimensional periodic table could be designed which includes all variables via pressure temperature and density v volume chaotic relationship. The latter would need to be interpolated. In any case such a venture if culminating in success would be valuable for future science and I would personally recommend the successful scientist/s for a Nobel prize nomination.

The other mass discrepancy has already been explained to be caused by space filling parameters of the nucleus and because H1 has no space filling problems its mass value can be declared to be almost exactly one. I say almost because the results are related to water which has an oxygen atom which has an unknown space filling arrangement of it nucleons. However it is an atom on the small side so the discrepancy may be in the order of many significant zeros after the decimal point.

There is another mass dilemma caused by the varying transitions of a finite number of gravitons per unit volume. It is possible that this may actually iron out some of the other discrepancies and the mass deficit may even become a credit in certain cases. All in all, this cannot be an accurate science until we can get inside nucleus, and the experimentally evaluated results are what we have to work with, and we simply have to explain the discrepancies away for the time being.

In this sense, the mole, atomic number and molar mass remain unchanged because they have been fairly accurately derived from observed experiments even if some inaccuracies still exist. All I have just done is address the AN-AM mass discrepancy non-relativistically. The previously noted C(mol) mass problem will remain but now for different reasons that have been tabled herein. I.e. being the degrees of freedom variability and the two mass discrepancy problems just analyzed.

If the Mass of Hydrogen is recalculated and the above methods used to calculate the atomic masses of all the other elements then the periodic table is "kaput". There is no valid reason to do that unless new science requires greater accuracy of the subjective terms of this analysis.

Science has been happily able to advance to welcome and laudable technologies to date, even with the discrepancies and many novel and plausible sounding reasons have been given to explain this. Interestingly the current quantum model seems to be a "box on wheels" just like mine. The question is: Which model can be "pimped" the best to answer the enigmas and contradictions which are comprehensively addressed and hopefully solved with this presentation?

You may leave me now long suffering reader, because here is where is where I get to be burnt at the stake!

A vibrating force with a linear motional component can transfer that component to another object within its field of influence and it can exhibit trans-dimensional effectiveness. All objects with atomic mass would by classical physics, exhibit a perfectly elastic collision process because every action has an equal and opposite reaction and whether it is an actual collision or a force collision the result must be the same. This means that unless miracles are occurring at an astonishing rate then the atomic objects must be seen to have no actual mass or even effective mass either for that matter either. (I know; another unintentional . . .)

So there must be a miracle occurring because inelastic collisions occur all the time everywhere. The miracle is performed by gluons and their sphere of influence which is the strong binding force and to a lesser extent the other forces at work.

In putting these processes to work we can see how a transiting graviton can trade both a partial velocity component and also a possible matter component. In the first instance a non-confined graviton trades linear spatial velocity for vibrational amplitude (velocity) during transitions. This results also in a forced and instantaneous linear velocity change in both of the gravitons which are transiting each other. If one is inside an atom then the atom will instantly exhibit a linear velocity change in proportion to the closing velocity and the binding force confinement within the atom which is caused by the biracial bonding forces which are holding it together, and the motional changes can be considered to be caused by force interactions and not by any perceived physical contact This would be impossible in the real world of classical physics but because of the forces being fundamentally caused by a cosmic affect of matter anti-matter attractional force it can and it does.

In AMOs the effect of a high volume of multilateral graviton transitions is transferred through the object the same way that any motion affecting force is. That is via the binding forces. Again there is never any actual physical contact of "hard" matter within any AMO or object in the universe. It is all held together and moved by FORCE! If only one graviton were to transit just one of the atoms in the AMO, the whole AMO would see the same velocity change as if it was only one atom. This is because AMOs have no MASS.

Transference of energy only occurs by the swapping of fundamental force particles which are able to be emitted by BBR or photonic radiation or by "handshake" swapping between atoms. If the gravitons only caused motional change (because they are massless particles) all that would be seen would be a straight trade of motion without any energy being transferred and used.

Any idea that mass converts to energy and the value can be determined by resultant velocities may be fine mathematically if assumptions are made regarding the terms of the formula. Notwithstanding this it is without feasibility, because there is no known mechanism in vector math which can account for a vibration amplitude change which has no linear component to cause a linear transfer of motional force with resultant momentum as an energy component.

This would also mean that at rest the object would have to have full energy and no mass and at "c" the reverse would be the case. This must then cause the assumption that we have the mass that we do because we are moving through space at about 30km/sec and everything including photons at the time of emission has less energy, and the interesting fact that the photon already has a velocity of 30km/sec. which would confer it with mass as well, and therefore instantaneous acceleration would be impossible.

This is very problematic for the supposed behavior of light in some theories. In objection to the idea that a photon has a vibrational amplitude velocity of "c" when at rest, and no vibration when traveling at "c" we should note the many different frequencies which are typical of various photons and ask for an explanation of how they can all be vibrating at "c". The infrared photon which has the lowest "energy" content would have to exhibit the greatest amplitude while an x-ray photon would have the lowest, and consequently the lowest "energy" state. Unfortunately for such a postulation, the reverse is seen to be true and it has no legs.

Now I am going to broach a subject with some philosophical overtones. This may appear to be like throwing salt on an open wound, but it must be said.

A whole arm of physics has grown up around $E=mc^2$. It requires that fermions have half integer spin and other particles to have spin to a varying degree in order to assure the conservation of energy through angular momentum. It all appears to sort of make sense, except that The electron volt (Ev) is actually the kinetic energy of an electron being forced to move by an emf (which is actually declared to not be a force after all) and this would then confer mass on the electron and not actually be a measurement of any energy. Somehow the force to energy transition comes easy when mind gaming the folks.

Calculations and assumptions of the Ev "energy" state of quantum and fundamental particles has been an evolution beginning with an arbitrarily connected relationship between electron motion and the supposed affect (by miraculous cause) on the size and spin moments of these particles and there appears to be the magical ability to call energy mass and mass energy at will. By my theory this whole idea is purely hypothetical and firmly ensconced in the miracle department because of the fact that it shows many anomalous and weird results by marrying weird physics with weird mathematics.

Of course the whole mish-mash is glossed over with a high degree of technical finesse and gouached over with a mind bogging array of formulae and technical diagrams and graphs.

I will give many scientists credit where it is due for stating the obvious and attempting to find answers to the many dilemmas evident in the postulations.

A real problem though is that once the energy values have been fudged to fit the formula E=mc^2 (which by the way can only be accurate in a real universe at zero k) the science can be seen to interrelate fairly comfortably albeit with artful rounding off and cheeky interpolations and this then allows it to evince credibility. Of course this might be vehemently denied by anyone who may not have studied the century long evolution of the science.

My main objection is that physicists have simply created a closed system with various laws, assumptions, mathematical formulae and algorithms that only apply to substantiate that system, and which can't really apply to the world of real physics which disallows objects to have no mass and even exist only as virtual reality!

Science is therefore seen to be caught in a bind of its own manufacture and the whole load needs to be dumped.

I will not put all scientists in the same basket but there will be some who will think that somehow their mathematical prowess is conclusive evidence in itself of the soundness of the theory. This is elitism of the worst kind and can only end in academic stalemate if someone doesn't cry out very loudly and often. "The Emperor has no clothes!!"

Of course this is going to be countered by . . . "But your theory contains miraculous and inexplicable forces and causations as well" I will readily admit that, but (in common with all theories), at the basement level there are the expected unknowns and many steer well clear of such a contentious area of thought because the idea of God comes to mind as the creator as well as the sustainer of the fundamental processes.

We all want to know the truth but many are afraid of where the search may end up. In any case how can we find the truth if we don't question existing theories with boldness?

EPILOGUE

In the end it comes down to either Einstein and Lorentz along with characteristically complex formulae and assumptive and intellectually twisted reasoning, or multi-dimensionalism and fundamental particle theory which allows standard scalar physics to operate in a standard and undistorted framework.

Relativity provided a stop-gap measure to enable some modicum o sense to be made of the nature of things, but it created its own mindset of "logical irrationality", a merism which by necessity had to be adopted in principle by most physicists by having to 'sell their soul" to it or else advance no further in the academy. As weird as multidimensionalism may appear; it does not contain logical dilemmas but it does have precedents as does particle theory, albeit often with relativistic overtones.

A point to note is: All that may be theorized in mathematics (being pretty much anything) does not necessarily translate into the observable universe. If it did, the universe would look nothing like it does in which a complex yet logical reality has arisen out of beautiful, (meaning ordered) chaos.

One should perhaps be mindful that the origin of the universe may indeed have been in some manner similar to my speculation. This would indicate that a far less complex and united cosmos resulted in a chaotic universe of even greater complexity and wonder, because of diversity. The reverse in chaos theory is impossible without specific laws and controlling algorithms!

As far as questioning what lays beyond the cosmos goes; it suggests speculations as unwieldy as the supposition of the existence of smaller sub particles of matter than sub-sub-particles, which are responsible for forces. To avoid questions of why rather than how, I feel that the subject should be contained between acceptable limits. Having said this I will go on the record (with all these things having been considered) and emphatically state that the graviton is the Higgs boson which has been speculated as being responsible for gravity and mass but not the "God particle". The God particle is a particle of no measurable size. It is small and large at the same time, and the whole universe finds its existence within it like a dilation particle. Yes the God particle is the Eos!

If scientists are searching for a "formula of everything" then I expect they will be very disappointed. Wouldn't it be better to realize that perhaps we would be better served by recognizing the overriding and hopefully altruistic dimension of the Theos?

These are questions that I have not attempted to answer. I'll leave that ball in your court.

1/ Does the Eos receive and transmit instantaneous data from and to somewhere or someone?

2/ How can an unintelligent Eos control the forces that create and sustain living beings?

3/ Who or what damaged the cosmos in the first place?

4/Could more gravitons come flooding into the universe from the cosmos and so upset the equilibrium in a short amount of time?

5/ Who . . . . ? "You fill in the gaps!"

So which is it to be; "movement at the station" to act upon this idea with scientific vigor, or will you continue waiting for a deeper level of non weird science?

RATIONALE AND CONCLUSIONS:

It may be beyond human mental capacity to fully understand the physics of the natural universe. Maybe we need some help from above!

While some of the ideas and theories within the main postulation of a multidimensional universe may not be as I have surmised. This would only lend difficulty to the theory if such incorrect assumptions could prove the general theory impossible.

Parts of this theory although wholly original in thought have already been theorized by other smarter men than me and I take no credit where none is due. All I have attempted to do is to solve a universal problem of lack of congruity at both the universal and quantum level to perhaps enable scientists to direct their attention back to possibilities such as some of those entertained in "Star Trek"!

It should by now appear that a seemingly chaotic universe was formed out of pre-existing order and structure, and so being held together by the original cosmic laws which are now less than capable of bringing about the return to order because of conflicting requirements. However; such happenstance is the fundamental reason that I am at all able to write this and that you are even alive to be able to read it.

The cosmic laws know nothing of a universal law which states that chaos cannot lead to order but only further chaos, so the cosmo-universal particles continue their futile quest unabated. This leads to the conclusion that forces at the fundamental level also arise from outside our own universe, and that they originated before its time.

The universe is destined to become just a scar in the hyperverse forever.

This theory also concludes that everything is made from the same sub-particles of matter. Black holes and quarks are cousins!

I would have wished for the theory to lead to a more positive future predication for the universal existence, and perhaps for some discovery of technological wonder. At the same time as seeming to have failed in both endeavors, I fear that I may have reached other conclusions that will not sit well with you. I don't like to have to say this "but that's just too bad, because scientific search for truth was never supposed to be outcome driven. We must go wherever it takes us, like it or not!"

Even though I have left some questions unanswered, I have attempted to answer the many enigmatic, contradictory and downright illogical ones to the end point which always has been threatening to certain sensitivities about the subject of origins. That is; that the real questions that remain are now philosophical and we are sailing very close to the wind that points to an intelligent and fundamentally very powerful creator.

With regard to an observer of no particular size, you may be tempted to "sub-let" space with smaller and smaller levels of particles ad infinitum. However there comes a point where there remains no further rational reason because there is no physical evidence or observational necessity for such postulation. This is the point at which we need to stop.

In this treatise I have come as close as I dare to the purely metaphysical, (and I may even be guilty of having crossed the line a little) for the following reasons.

We arrive at a point where repulsive, attractive, additive subtractive, color, strangeness, etc forces must be deemed to be "felt" observances. That point I believe is the realm of biracial, electrostatic, magnetic and other dimensional forcefields which occur below levels at which particle motion can be observed.

The sub-particles are moving by elastic interaction and positioning due to "energy" state requirements and it is the felt operation of the sum of their motional parts which results in what we observe as virtual force/energy. Such activity can be notably observed in the operation of an electrical transformer in which there are no externally moving parts, but the internal motion of force particles creates the illusion of force by subsequent variance in their field states, so causing energy transfer without the transfer of the force itself.

This is not actually energy transfer. The magnetic force acts like a catalyst to convert force particle motion in one conductor to enforce particle motion in the other conductor by proximity field affects, and it is only when this causes actual work to be done (such as light up a light bulb) that energy is seen to be released.

My postulation of the Eos being a single particle of indeterminate size is with reference to the "no size" observer; and it is necessary because of the requirement for there to be instantaneous force balance throughout the universe because of its inelasticity.* This is not equivalent to energy balance which is transmitted via rate of divergence. If this cannot be the explanation then metaphysical explanations of universal operation become valid. *Within the constraints that I have previously mentioned.

I must admit that the idea of a separation force or solid matter repulsive force at the extra-near-field level* has no known cause except for gluon drought. However it lends itself to actual realization because of the fact that similar elemental solids do not weld together on contact. This force is deemed to be a parameter of the Eos. It is highly likely that the reason actually is; that there is a "drought" of gluons on the surface of every solid object for reasons which I will leave for now because I don't have the mental energy left to write a "two squillion" page book! Even though there seems to be a universe of even tinier particles inside of atoms and atomic particles. I'm just happy with theorizing about the "galaxy" sized objects that are in there. *At or just beyond the outer electron orbital.

There is therefore no need for visualizing any ultra or parallel universes, or realties. Such things are only stuff of the imagination and there remains nothing that I know of which is observed in the physical world that requires their existence. This becomes especially so if closed circular event paths, able to result in an infinite number of possible eventualities can be reasoned.

This in no way suggests that any life form or natural event can have any other possible eventuality than the ones which are being realized and which have also been experienced in the past. Any other possible hope for future existence in other realms or universes is a matter for faith. This is because of the legal unreasonableness of allowing other existences to be postulated outside of the cosmo/universal laws which operate in this universe. Such mental constructs cannot be proven. The fact that we are alive perhaps, should give us moment to pause!

There is nothing that prevents the postulation of other mechanical and mental constructs which (however cozy and warm they may seem), may nevertheless be untenable from the point of view of empirical science.

We could by mental construction imagine electric, magnetic and other fields existing by the motion of multiple sub-particles traveling via their own field lines in order to propagate the "felt" forces which have been realized, but by similar scientific reasoning, there is no evidence for such speculation to be required, and neither does there remain any scientific or mathematical dilemma which requires the postulation of such. In my opinion it remains unnecessary and is therefore not concluded to be part of this theory.

In brainstorming such a "maniacally" radical model, I was not overcome by the desire to be smart or cute in any intellectually superior manner. I developed most ideas from the theoretical models of others and began to see a way of unifying them in a method which could reasonably overcome the many difficulties which are recognized in currently accepted models of classical and theoretical physics and more so in quantum physics and cosmology.

By now you should realize that the arbitrary conference of Ev to either energy mass is a "no brainer", and that the whole of the last hundred years of quantum and relativistic postulation will have to be discarded!

In saying that; I recognize the massive amount of knowledge that has been gained by science in that time. The only problem has been in tying up the loose ends. It may seem too daunting a challenge to make the switch to multi-dimensionalism, and so if science and society are happy with their lot, then I will conform: "Leave well enough alone".

At first I attempted to develop a postulation which utilized wave theory, but I quickly realized that this could in no way fit with observance of the natural universe and so I decided to take an approach which was a total (if not obvious) paradigm shift over to particle theory, just to check for a reasonable fit. In so doing, things began to fall into place in (as now recognizably) amazing ways.

The universe appears to have arisen out of a cataclysmic shattering of the original cosmos, which has ever since, been constantly engaged in a vain attempt to restore itself. This attempt has been slowed down by the time delayed interaction of the damaged yet "quaint" cosmo-universal matter. This has occurred for some time and to such an extent that the universe has reached observable (short observational time period) equilibrium.

This equilibrium, which allowed the flourishing of life on Earth, may be shorter-lived than we expect. Global warming (as dire as it may seem) pales into insignificance when compared to the possibility of universal cooling and the "lights going out for good"! By way of reaction to this, you may reach inside for a little mirth, but that may well be cold comfort as you begin to realize that what you are observing with your telescopes is actually the far distant past, and that even now the objects you can "see" shining in the night sky may well be cold and extinct, even as you read this.

This leaves little to the imagination, insofar that even the most mentally challenged amongst us would realize that unless other avenues for escape should open, then the human race is probably doomed far sooner than "astro-intellectual" scientists would have you believe, and no matter what other scientific advances may be made for our comfort, and no matter what humanitarian ventures are engaged in, the only hope for a humanity that cannot "get the heck out of here" by themselves, is to seek help from outside the universe. Logic declares that if the universe exists then it exists within or perhaps even parallel with some other non cosmo-universal domain, and these are my final words.

www.ingramcontent.com/pod-product-compliance
Lightning Source LLC
Chambersburg PA
CBHW030005190526
45157CB00014B/435